CALVERT
MATH

Practice and Enrichment Workbook

CALVERT® SCHOOL

Calvert Math is based upon a previously published textbook series. Calvert School has customized the textbooks using the mathematical principles developed by the original authors. Calvert School wishes to thank the authors for their cooperation. They are:

Audrey V. Buffington
Mathematics Teacher
Wayland Public Schools
Wayland, Massachusetts

Alice R. Garr
Mathematics Department Chairperson
Herricks Middle School
Albertson, New York

Jay Graening
Professor of Mathematics
    and Secondary Education
University of Arkansas
Fayetteville, Arkansas

Philip P. Halloran
Professor,
Mathematical Sciences
Central Connecticut State University
New Britain, Connecticut

Michael Mahaffey
Associate Professor,
    Mathematics Education
University of Georgia
Athens, Georgia

Mary A. O'Neal
Mathematics Laboratory Teacher
Brentwood Unified Science
    Magnet School
Los Angeles, California

John H. Stoeckinger
Mathematics Department Chairperson
Carmel High School
Carmel, Indiana

Glen Vannatta
Former Mathematics Supervisor
Special Mathematics Consultant
Indianapolis Public Schools
Indianapolis, Indiana

CHIEF LEARNING OFFICER

Gloria Julius, Ed.D.

CURRICULUM CONSULTANT

Robin A. Fiastro, *Manager of Curriculum*

PROJECT FACILITATORS/LEAD RESEARCHERS

Hannah Miriam Delventhal, *Math Curriculum Specialist*
Patricia D. Nordstrom, *Math Curriculum Specialist*

SENIOR CONSULTANT/PROJECT COORDINATOR

Jessie C. Sweeley, *Associate Director of Research and Design*

WORKBOOK AUTHORS

Kathy S. Agley	Vicki Dabrowka	Gillian M. Lucas	Laura A. Rice
G. Louise Catlin	Carol A. Huggins	Patricia D. Nordstrom	Victoria B. Strand
Shannon E. Cheston			

EDITORS

Anika Trahan, *Managing Editor*

Eileen Baylus	Anne E. Bozievich	Danica K. Crittenden	Mary E. Pfeiffer
Holly S. Bohart	Maria R. Casoli	Pamela J. Eisenberg	Megan L. Snyder

GRAPHIC DESIGNERS

Lauren Loran, *Manager of Graphic Design*

Caitlin Rose Brown	Vickie Johnson	Deborah A. Sharpe	Teresa M. Tirabassi
Steven M. Burke	Bonnie R. Kitko	Joshua C. Steves	

For information regarding the CPSIA on this printed material call 203-595-3636 and provide reference # RICH - 312337-3W

ISBN-13: 978-1-888287-73-8

Contents

Name **Dalaney** JONES

Number Patterns Using a Number Grid

600	601	602	603	604	605	606	607	608	609
610	611	612	613	614	615	616	617	618	619
620	621	622	623	624	625	626	627	628	629
630	631	632	633	634	635	636	637	638	639
640	641	642	643	644	645	646	647	648	649
650	651	652	653	654	655	654	657	658	659

Use the number grid to complete problems 1–3.

1. Ring all numbers with 7 in the ones place.

2. Mark X on all numbers with 2 in the tens place.

3. Draw □ around the number that is both ringed and marked X.

Count by tens. Write the missing numbers.

4. 54, 64, __74__

5. __816__, 826, 836

6. 172, __182__, 192

7. 430, 440, __450__

8. __705__, 715, 725

9. 587, __597__, 607

Count by hundreds. Write the missing numbers.

10. 324, 424, __524__

11. 645, __745__, 845

12. __39__, 139, 239

13. 200, __300__, 400

14. __326__, 426, 526

15. __772__, 872, 972

Solve. Remember to label your answers.

16. Laura did 24 sit-ups in a row. If Jen did 10 more sit-ups than Laura, how many sit-ups did Jen do?

    _24 + 10 = 34_

17. Bill is thinking of a number. The number is 100 less than 596. What is Bill's number? _596 − 100 = 496_

MIXED Practice

Add or subtract.

1. $\begin{array}{r} 6 \\ + 7 \\ \hline 13 \end{array}$

   2. $\begin{array}{r} 9 \\ + 3 \\ \hline 12 \end{array}$

   3. $\begin{array}{r} 8 \\ + 6 \\ \hline 14 \end{array}$

   4. $\begin{array}{r} 7 \\ + 7 \\ \hline 14 \end{array}$

5. $\begin{array}{r} 12 \\ - 6 \\ \hline 6 \end{array}$

   6. $\begin{array}{r} 12 \\ - 6 \\ \hline 0 \end{array}$

   7. $\begin{array}{r} 12 \\ - 6 \\ \hline 6 \end{array}$

   8. $\begin{array}{r} 16 \\ - 9 \\ \hline 7 \end{array}$

Write the missing numbers.

9. $10 = \boxed{7} + 3$

10. $2 + \boxed{9} = 11$

11. $6 = 6 + \boxed{0}$

12. $14 - \boxed{10} = 4$

13. $9 - 6 = \boxed{3}$

14. $10 = 2 + \boxed{8}$

Name _____

Place Value to Thousands

Write each number.

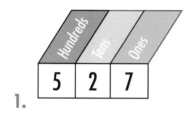

1. _____

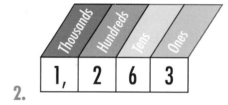

2. _____

3. _____

4. _____

Write the digit in each place value.

5. 635 = _____ hundreds _____ tens _____ ones

6. 451 = _____ hundreds _____ tens _____ ones

7. 704 = _____ hundreds _____ tens _____ ones

8. 2,946 = _____ thousands _____ hundreds _____ tens _____ ones

9. 8,327 = _____ thousands _____ hundreds _____ tens _____ ones

10. 4,025 = _____ thousands _____ hundreds _____ tens _____ ones

Write each number.

11. three hundred seventy-two _____

12. five thousand, two hundred nine _____

13. four hundred eighty-nine _____

14. eight thousand, six hundred forty _____

Write the place value of the 6 in each number.

15. 3,624 _____   16. 687 _____   17. 7,306 _____

18. 6,200 _____   19. 1,560 _____   20. 465 _____

Solve. Remember to label your answers.

21. Tom's street address is six thousand, four hundred thirty-seven Main Street. What digit is in the tens place of Tom's street address?

    _____

22. Write the number that has a 3 in the hundreds place, an 8 in the ones place, a 4 in the thousands place, and a 2 in the tens place.

    _____

23. Stamps are sold in books of 10. In one week, Mrs. Smith used 5 books and 3 stamps from another book. How many stamps did she use in all?

    _____

Name _____

Place Value to Thousands

Write the least and greatest 4-digit number that has a...

1. 4 in the hundreds place.   least ___1,400___
                                 greatest ___9,499___

2. 7 in the ones place.   least _____
                                 greatest _____

3. 5 in the thousands place.   least _____
                                 greatest _____

4. 1 in the ones place.   least _____
                                 greatest _____

5. 9 in the hundreds place.   least _____
                                 greatest _____

To add 100 to a number using mental math, increase the digit in the hundreds place by 1. 4,826 + 100 = 4,926

Add 100 to each number using mental math.

6. 7,320 _____   7. 9,654 _____   8. 2,015 _____

To add 1,000 to a number using mental math, increase the digit in the thousands place by 1. 6,218 + 1,000 = 7,218

Add 1,000 to each number using mental math.

9. 5,248 _____   10. 7,054 _____   11. 2,879 _____

Subtract 100 from each number using mental math.

12. 2,873 _____   13. 1,450 _____   14. 8,261 _____

Name _____

Standard and Expanded Form

Fill in the place-value chart. Then read the number aloud.

1.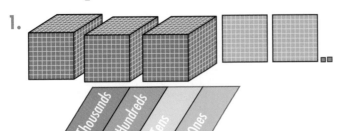

2.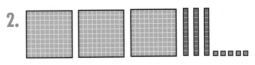

3.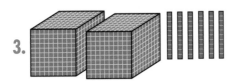

4.

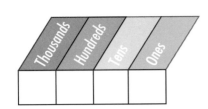

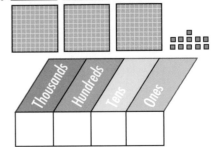

Write these numbers in expanded form. Then read them aloud.

5. 3,256 = _____ + _____ + _____ + _____

6. 740 = _____ + _____

Write these numbers in standard form and word form.
Then read them aloud.

7. 700 + 40 + 6 = _____

8. 6,000 + 300 + 20 + 9 = _____

_____

_____

9. 4,000 + 20 + 5 = _____

10. 900 + 7 = _____

_____

_____

Name _____

Greater Numbers through Hundred Thousands

Write the numbers in the place-value chart. Then read them aloud.

Example: 27,400

1. 9,256
2. 76,803
3. 82,045
4. 394,700
5. 608,401
6. 715,829

Hundred Thousands	Ten Thousands	Thousands	Hundreds	Tens	Ones
	2	7,	4	0	0

Write each number in standard form and expanded form.

7. fifty-three thousand _____

   _____ + _____

8. forty-seven thousand, eight hundred twenty-four _____

   _____ + _____ + _____ + _____ + _____

9. two hundred twenty thousand, six hundred seventy _____

   _____ + _____ + _____ + _____

10. nine hundred seventy-six thousand, three hundred thirteen _____

    _____ + _____ + _____ + _____ + _____ + _____

Solve. Remember to label your answers.

11. On the Fourth of July, seven thousand, nine hundred thirty people attended a baseball game. Write the number in standard form.

   _____

12. The distance around Earth at the equator is twenty-four thousand, nine hundred one miles. Write the number in standard form.

   _____

13. One year on the planet Saturn is equal to ten thousand, seven hundred fifty-nine Earth days. Write the number in standard form.

   _____

14. Use the digits 0, 1, 2, 4, 6, and 7 to write the least possible 6-digit number.

   _____

MIXED Practice

Write the missing number.

1. $6 + \boxed{} = 14$

2. $8 + 6 = \boxed{}$

3. $14 - \boxed{} = 6$

4. $14 - \boxed{} = 8$

5. $4 + 2 + 8 = \boxed{}$

6. $6 + 2 + 7 = \boxed{}$

Name _____

Place Value

Write the number.

1. I am a 5-digit number less than 10,001.

   _____

2. I am 500 less than the smallest 4-digit number with 9 hundreds.

   _____

3. I am a 5-digit number with 4 ten-thousands and 4 ones. My other digits are 7s.

   _____

4. I am 200 less than the largest 4-digit number.

   _____

5. I am the smallest 4-digit number with 3 hundreds and 9 ones.

   _____

6. I am the largest 4-digit number with 6 tens and 2 ones.

   _____

7. I am the smallest 5-digit number with 0 thousands and 5 tens.

   _____

8. I am 10 less than the smallest 4-digit number.

   _____

Name _____

Even and Odd Numbers

Even numbers always end with 0, 2, 4, 6, or 8, and odd numbers always end with 1, 3, 5, 7, or 9.

Write even or odd.

1. 83 _____   **2.** 94 _____   **3.** 350 _____   **4.** 77,392 _____

5. 408 _____   **6.** 52,961 _____   **7.** 1,423 _____   **8.** 500 _____

Solve.

9. Write a number with 7 in the thousands place, 3 in the hundreds place, and 6 in the tens place. In the ones place, choose and write any digit that makes the number an odd number. Read the number aloud.

_____

10. Write the even number that is greater than 324 but less than 327.

_____

Write the missing numbers. Then answer the questions.

11. 73, 75, 77, _____, _____, _____

Did you write even or odd numbers? _____

12. 502, 522, _____, _____, _____

Did you write even or odd numbers? _____

13. 4,209; 4,211; _____; _____; _____

Did you write even or odd numbers? _____

Solve.

14. If you add an even and an odd number, will the sum be even or odd? Explain.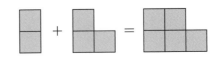

    _____

    _____

15. If you subtract an even number and an odd number, will the difference be even or odd? Draw a sketch and explain.

    _____

    _____

16. Lucia's town is trying to save water. Residents who have even-numbered street addresses may only water their lawns on even-numbered days. If Lucia's address is 897 Elm Street, can she water her lawn on July 6? Explain your answer.

    _____

    _____

17. During a music program, the third graders walk into the auditorium in pairs. If there are 37 third graders, will everyone have a partner? Explain your answer.

    _____

    _____

Name _____

Writing a Constructed Response

Solve. Remember to label your answers.

1. In their books, Tom read 84 pages, John read 101 pages, and Dave read 79 pages. Who read the greatest number of pages? Who read the least number of pages? Explain how you know.

   _____

   _____

   _____

2. The number 79,863 is increased by 10,000. Write the new number in standard form. Explain how the original number was changed.

   _____

   _____

   _____

3. Mr. Chen has $34,528 in his bank account. He is going to deposit $6,000 into his account and give $10,000 to his favorite charity. How can Mr. Chen use place value and mental math to figure out how much money he will have left in his account?

   _____

   _____

   _____

4. Eric says to Becky, "I am thinking of a number with zero hundreds, five ones, three ten thousands, six tens, and eight thousands. What number is one hundred more than my number?" Explain how Becky can find Eric's number.

_____

_____

_____

5. One of the numbers on Frank's mailbox fell off. It is greater than the ones digit, but less than the hundreds digit. It is an odd number. What number fell off? Explain how you know.

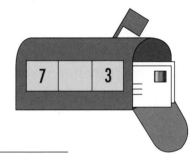

_____

_____

_____

6. Which is more, three tens or four ones? Explain how you know.

_____

_____

_____

Name _____

Comparing Numbers

Remember:
< means is less than
> means is greater than
= means is equal to

Hint: The comparison sign always points to the lesser number.

Ring the greater number.

1. 439	2. 809	3. 26,527	4. 65,412	5. 928,967
430	890	26,745	63,432	928,947

Compare. Write <, >, or =.

6. 7,321 ◯ 7,421   7. 929 ◯ 919   8. 515,505 ◯ 515,515

9. 43,888 ◯ 43,828   10. 15,230 ◯ 16,230   11. 9,403 ◯ 9,403

12. 752 ◯ 742   13. 94 ◯ 104   14. 632,860 ◯ 632,850

Solve. Remember to label your answers.

15. Matt swam underwater for 35 seconds. Kevin swam underwater for 38 seconds. Who swam underwater longer? Explain your answer.

_____

_____

16. In a jump rope contest, Kelly made 573 jumps before she missed. Kurt was able to jump 571 times before he missed. Who made more jumps, Kelly or Kurt? Explain your answer.

_____

Name _____

Ordering Numbers

Order the numbers from least to greatest.

1. 39, 85, 84 _____ , _____ , _____

2. 7,310; 7,301; 7,103 _____ ; _____ ; _____

3. 15,921; 51,129; 15,219 _____ ; _____ ; _____

4. 62,398; 58,239; 68,293 _____ ; _____ ; _____

Order the numbers from greatest to least.

5. 821, 812, 802 _____ , _____ , _____

6. 4,320; 4,203; 4,230 _____ ; _____ ; _____

7. 24,851; 42,158; 42,518 _____ ; _____ ; _____

8. 63,954; 63,459; 63,594 _____ ; _____ ; _____

Solve. Remember to label your answers.

9. Tomás wrote the following numbers: 27,417; 27,565; 27,893. He said the numbers were in order from greatest to least. Is Tomás correct? Explain.

   _____

Name _____

Rounding Numbers to Ten Thousands

Write the ten that comes before and after each number.
Ring the ten that is closer to the given number.

1. _____ 63 _____   2. _____ 48 _____   3. _____ 17 _____

Round to the nearest ten.

4. 18 _____   5. 43 _____   6. 25 _____   7. 41 _____

Use the number line to round to the nearest hundred.

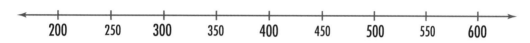

8. 375 _____   9. 415 _____   10. 528 _____   11. 250 _____

Write the hundred that comes before and after each number.
Ring the hundred that is closer to the given number.

12. _____ 487 _____   13. _____ 536 _____   14. _____ 223 _____

Round to the nearest thousand.

15. 5,285 _____   16. 3,405 _____

17. 6,850 _____   18. 9,538 _____

Round to the nearest ten thousand.

19. 33,115 _____   20. 57,863 _____   21. 85,724 _____

Name _____

Rounding

These numbers are rounded to the nearest ten.

6⃝3 9 ⟶ 640   7⃝2 1 ⟶ 720   8⃝2 5 ⟶ 830

↑   ↑
tens   1 zero

These numbers are rounded to the nearest hundred.

5,⃝8 76 ⟶ 5,900   64,⃝2 29 ⟶ 64,200   86,⃝5 91 ⟶ 86,600

↑   ↑
hundreds   2 zeros

These numbers are rounded to the nearest thousand.

⃝8 ,470 ⟶ 8,000   5⃝6 ,851 ⟶ 57,000   1⃝8 ,501 ⟶ 19,000

↑   ↑
thousands   3 zeros

The ringed numbers show to which place you are rounding. Look at the digit to the right of the ringed number. If that digit is 5 or greater, the ringed number increases by 1. If that digit is less than 5, the ringed number remains the same. All digits after the ringed digit become zeros.

Round to the nearest ten.

1. 7⃝3 6 _____   2. 4⃝5 3 _____   3. 8⃝7 5 _____

Round to the nearest hundred.

4. ⃝5 49 _____   5. 7,⃝2 05 _____   6. 2,⃝7 61 _____

Round to the nearest thousand.

7. 5⃝6 ,534 _____   8. 47⃝6 ,125 _____

Name _____

Problem-Solving Skill: Using the Four-Step Plan

Solve.

1. Emma has 362 paper clips. Grant has 392 paper clips. Carson has 403 paper clips. Who can build the longest paper clip chain? Explain your answer.

    _____

    _____

2. I am thinking of a mystery number. If 1,000 is added to the number, the total is 1,475. What is the mystery number? Explain how you know.

    _____

    _____

3. Darlene wrote a number. The number has 1 ten. The number rounded to the nearest ten is 20. What numbers could Darlene have written?

    _____

    _____

4. What are the least and greatest numbers that round to 400 when rounded to the nearest hundred?

    _____

    _____

5. Use the chart to order the depths from least to greatest.

_____; _____; _____; _____

Ocean Depths

Name	Where You Find It	Depth
Puerto Rico Trench	Atlantic Ocean	28,374 ft
Arctic Basin	Arctic Ocean	17,881 ft
Mariana Trench	Pacific Ocean	36,201 ft
Java Trench	Indian Ocean	25,344 ft

MIXED Practice

Add.

1. 8
  + 6

2. 4
  + 7

3. 9
  + 2

4. 8
  + 5

5. 4
  + 4

6. 8 + 4 = _____

7. _____ = 6 + 7

8. 4 + 1 + 8 = _____

Subtract.

9. 17
  − 8

10. 13
  − 5

11. 12
  − 6

12. 18
  − 9

13. 15
  − 9

14. 17 − 9 = _____

15. 16 − 7 = _____

16. _____ = 15 − 8

Write the missing numbers.

17. ___ + 3 = 11   8 + 3 = ___

18. ___ − 9 = 8   ___ − 8 = 9

11 − ___ = 8   11 − ___ = 3   9 + 8 = ___   8 + 9 = ___

Name _____

Chapter 1 Review

Write *true* or *false*.

1. In 9,382 the 3 has a value of 300. _____

2. Five hundreds, three tens, and seven ones are written as 5,307. _____

3. 898 is greater than 889. _____

4. Ten tens equal one hundred. _____

5. Ten hundreds equal one hundred thousand. _____

Write the sign for each phrase.

6. is greater than _____   7. is less than _____   8. is equal to _____

Write the number.

9. 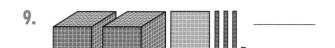 _____   10.  _____

11. 4 tens 6 ones _____   12. 8 hundreds, 3 tens, 7 ones _____

13. sixty-five _____   14. four thousand, seven hundred _____

15. 3 thousands, 2 tens _____   16. twenty-five thousand _____

17. 6 hundred thousands, 5 tens, 1 one _____

18. 3 ten thousands, 2 hundreds, 8 tens _____

19. 9,000 + 50 + 4 _____

© Calvert School

Write the number of hundred thousands, ten thousands, thousands, hundreds, tens, and ones.

		Hundred Thousands	Ten Thousands	Thousands	Hundreds	Tens	Ones
20.	563						
21.	6,048						
22.	35,709						
23.	402,195						

Write the missing numbers.

24. 47, 48, _____, _____, 51   **25.** 673, 674, _____, _____, 677

26. 428, _____, 448, 458, _____   **27.** 373, 473, 573, _____, _____

Write *even* or *odd*.

28. 407 _____   **29.** 8,485 _____   **30.** 300 _____

Compare. Write <, >, or =.

31. 289 ◯ 829   **32.** 568 ◯ 568   **33.** 5 tens ◯ 7 tens

Order these numbers from least to greatest.

34. 762, 267, 627   **35.** 8,912; 8,921; 8,291

_____, _____, _____   _____, _____, _____

Round to the nearest ten thousand.

36. 31,245   **37.** 97,531   **38.** 16,450   **39.** 26,855

_____   _____   _____   _____

Round to the nearest thousand.

40. 3,625 _____   **41.** 2,287 _____   **42.** 48,925 _____   **43.** 7,556 _____

Name _____

Chapter 1 Test Prep

Ring the correct answer.

1. What number comes next? 329, 429, 529, _____
    a. 530
    b. 600
    c. 629
    d. 679

2. Which number shows *twenty-eight thousand, four hundred sixty-three*?
    a. 2,863
    b. 28,463
    c. 28,643
    d. 280,463

3. What is the place value of the 3 in the number 58,396?
    a. ten thousands
    b. thousands
    c. hundreds
    d. tens

4. Which shows 820,506 written in expanded form?
    a. 8,000 + 200 + 50 + 6
    b. 800,000 + 2,000 + 500 + 6
    c. 80,000 + 2,000 + 50 + 6
    d. 800,000 + 20,000 + 500 + 6

5. What number is missing from the pattern? 333, 335, ___, 339, 341
    a. 338
    b. 337
    c. 336
    d. 37

6. Which number is 100 greater than 5,836?
    a. 5,736
    b. 5,846
    c. 5,936
    d. 6,836

7. Which comparison is true?
    a. 6,845 > 6,475
    b. 393 < 390
    c. 51,492 > 52,249
    d. 702,013 > 702,031

8. Which group of numbers is ordered from least to greatest?
    a. 34,326; 34,427; 34,436; 34,508
    b. 418, 408, 438, 421
    c. 78, 65, 83, 59
    d. 6,602; 6,226; 6,662; 6,266

9. What number do these blocks represent?
    a. 35,540
    b. 3,540
    c. 3,504
    d. 354

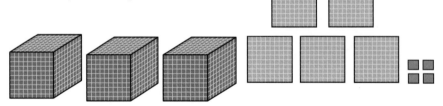

10. What number is 10 less than the smallest 4-digit number?
    a. 1,010
    b. 999
    c. 990
    d. 900

Name _____

Addition Strategies

Add. Write the letter of the addition strategy you used to find the sum.

_____ **1.** 2 + 8 = _____
_____ **2.** 4 + 5 = _____
_____ **3.** 9 + 3 = _____
_____ **4.** 3 + 7 = _____
_____ **5.** 9 + 2 = _____
_____ **6.** 5 + 6 = _____

Addition Strategies	
A. Count on	**B. Sums of 10**
3 + 8 = _____	5 + 7 = _____
Start with 8.	5 + 5 = 10
Count on 3 more.	7 is 2 more than 5
8: 9, 10, 11	5 + 5 + 2 = 10 + 2
3 + 8 = 11	5 + 7 = 12

Solve. Remember to label your answers.

7. In Randy's room, there are 8 shelves of books and 3 shelves of trains. How many shelves are in Randy's room?

_____

8. Explain how counting on can help you add 39 + 3.

_____

_____

9. Delilah knows that 5 + 5 = 10. How can this help her add 5 + 6?

_____

_____

Name _____

More Addition Strategies

Find each sum.

1. 9 + 4 = _____   2. 9 + 8 = _____   3. 6 + 6 = _____

4. 6 + 7 = _____   5. 7 + 9 = _____   6. 4 + 4 = _____

7. _____ = 2 + 8   8. _____ = 4 + 5   9. 9 + 9 = _____

10. 1 + 11 = _____   11. _____ = 7 + 7   12. 8 + 6 = _____

Find each sum. Look for a pattern in each row.

13. 8 + 7 = _____   8 + 17 = _____   8 + 27 = _____   8 + 37 = _____

14. 4 + 6 = _____   14 + 6 = _____   24 + 6 = _____   34 + 6 = _____

15. 8 + 5 = _____   18 + 5 = _____   38 + 5 = _____   58 + 5 = _____

16. 4 + 9 = _____   4 + 19 = _____   14 + 19 = _____   14 + 29 = _____

Solve. Remember to label your answers.

17. Keith drew 9 pictures of people and 6 pictures of animals. How many pictures did Keith draw altogether?

_____

18. Gordon bought 8 green picture albums and 8 blue picture albums. How many picture albums did Gordon buy in all?

_____

Name _____

Fact Families

Complete each fact family.

1. $6 + 7 = 13$ **2.** $3 + 5 = 8$ **3.** $8 - 7 = 1$

_____   _____   _____

_____   _____   _____

_____   _____   _____

Use each group of numbers to write a fact family.

4. 7, 9, 16   **5.** 9, 8, 17   **6.** 7, 8, 15

_____   _____   _____

_____   _____   _____

_____   _____   _____

_____   _____   _____

Find the missing number to make each equation true.

7. $3 + 8 = \boxed{}$ **8.** $\boxed{} = 15 - 6$ **9.** $7 + 6 = \boxed{}$

10. $\boxed{} = 12 - 5$ **11.** $3 + \boxed{} = 12$ **12.** $4 + \boxed{} = 12$

13. $13 - \boxed{} = 8$ **14.** $\boxed{} - 7 = 7$ **15.** $10 = \boxed{} + 9$

16. $\boxed{} = 4 + 7$ **17.** $6 = 4 + \boxed{}$ **18.** $8 + \boxed{} = 8$

Name _____

Associative Property

Underline the numbers you are to add first. Then add.

1. $5 + (1 + 9) =$ _____

2. $(4 + 6) + 8 =$ _____

3. $5 + (3 + 7) =$ _____

4. $(8 + 2) + 2 =$ _____

Add. Hint: Look for the sum of 10.

5.	6.	7.	8.	9.
6	6	5	7	1
4	9	8	8	6
+ 9	+ 1	+ 5	+ 3	+ 4

Solve. Remember to label your answers.

10. Bert caught 7 fish one day, 5 fish the next day, and 5 fish the third day. How many fish did he catch altogether?

_____

11. Amy has 9 guppies, 4 angelfish, and 7 black mollies in her aquarium. How many fish in all are in her aquarium altogether?

_____

12. Tomika saved $8 in January, $4 in February, and $2 in March. How much money did Tomika save in all?

_____

MIXED Practice

1. Ring the odd numbers.

   2 5 6 3 8 9 12 15

   24 35 56 89 45 76

2. Ring the even numbers.

   247 612 588 873 2,345

   5,854   2,346   6,543

Name _____

Comparing Addition Expressions

Compare. Write <, >, or =.

1. $3 + 9 \bigcirc 4 + 8$ 2. $5 + 5 \bigcirc 6 + 6$ 3. $7 + 6 \bigcirc 4 + 9$

4. $8 + 5 \bigcirc 2 + 2$ 5. $1 + 9 \bigcirc 3 + 8$ 6. $4 + 2 \bigcirc 5 + 1$

7. $3 + 1 + 5 \bigcirc 2 + 8 + 2$ 8. $5 + 4 + 3 \bigcirc 2 + 4 + 6$

Solve. Remember to label your answers.

9. Belle and Sarah planted flowers in their gardens. Belle planted 3 daisies, 6 tulips, and 4 violets. Sarah planted 5 sunflowers, 2 lilies, and 8 marigolds. Who planted more flowers?

_____

10. Johnny and Jamie each have a collection of marbles. Johnny has 10 blue marbles, 8 red marbles, and 2 tiger eye marbles. Jamie has 9 green marbles, 3 purple marbles, and 5 rainbow marbles. Who has more marbles?

_____

11. The zookeeper feeds 4 elephants, 3 giraffes, and 5 hippos on Monday. On Tuesday, he feeds 4 seals, 6 penguins, and 8 alligators. On which day does he feed more animals?

_____

Name _____

Properties of Addition

Commutative Property	Associative Property	Identity Property
Changing the order of the addends does not change the sum. 35 + 5 = 40 5 + 35 = 40	Changing the grouping of the addends does not change the sum. (6 + 4) + 8 = 18 6 + (4 + 8) = 18	When 0 is added to a number the sum is that number. 197 + 0 = 197 0 + 4,072 = 4,072

Name the property shown. Write *Commutative Property,*
Associative Property, **or** *Identity Property.*

1. 18 + 2 = 2 + 18 _____

2. 82 + 0 = 82 _____

3. (9 + 5) + 6 = 9 + (5 + 6) _____

Add.

4.   8   5   5.   7   4   6.   5   9
    + 5   + 8   + 4   + 7   + 9   + 5
    ___   ___   ___   ___   ___   ___

7. 4 + (3 + 3) = _____   8. (8 + 1) + 5 = _____

9.   6   10.   7   11.   2   12.   5   13.   1
     3   6   1   8   9
    + 4   + 5   + 8   + 0   + 6
    ___   ___   ___   ___   ___

Solve. Remember to label your answer.

14. At the Decatur Zoo, Ted saw 17 chimpanzees and 0 gorillas in the primate house. How many chimpanzees and gorillas did Ted see altogether?

_____

Name _____

Estimating Sums by Rounding

Round to the nearest ten.

1. 63 _____   2. 29 _____   3. 41 _____   4. 49 _____

Round to the nearest hundred.

5. 701 _____   6. 876 _____   7. 205 _____

Round to the nearest thousand.

8. 5,511 _____   9. 4,987 _____   10. 3,204 _____

Estimate each sum by rounding to the greatest shared place value.

11.   280
   + 450   + _____

12.   75
   + 64   + _____

13.   2,999
  + 3,587   + _____

14.   2,034
  + 8,605   + _____

15.   79
  + 82   + _____

16.   109
  + 903   + _____

Solve.

17. A bakery sold 133 cookies on Saturday and 103 cookies on Sunday. About how many cookies were sold in all? _____

18. Write two numbers that, when rounded to the nearest ten, have a sum of 90. _____ _____

Name _____

Rounding as a Check for Accuracy

When a problem *does* require an exact answer, rounding can be used to check for accuracy.

Example: 31 + 157 = _____

First, round the addends to the greatest shared place value and find the estimated sum.

$30 + 160 = 190$

Next, find the exact sum.

$31 + 157 = 188$

Then, compare the estimate with the exact sum.

190 and 188 are close. The answer is probably correct.

If the estimate and the exact answer are *not* close, you may have made a mistake. Look back at your work, and fix any errors.

Estimate by rounding to the greatest shared place value. Then find the exact sum. If the answers are not close, try again.

1. 81
    + 18   + _____

2. 4,061
    + 238   + _____

3. 42
    + 54   + _____

4. 7,505
    +   382   + _____

5. 408
    + 501   + _____

6. 3,807
    + 4,191   + _____

Name _____

Problem-Solving Strategy: Drawing a Picture

Draw a picture to solve each problem. Remember to label your answers.

1. Emily works at a flower shop. She needs to fill 8 pots with 3 daisies each. She then needs to fill 2 pots with 2 tulips each. How many pots does Emily need altogether? How many flowers will she plant in all?

   _____

2. Margaret is going to knit mittens for her friends. She will knit 5 pairs of rainbow mittens, 5 pairs of blue mittens, 5 pairs of green mittens, and 5 pairs of red mittens. How many single mittens will Margaret knit in all?

   _____

3. Peter's striped cat had 8 striped kittens and his orange cat had 6 orange kittens. How many cats and kittens are there altogether?

   _____

Name _____

Adding by Regrouping Ones

Add. Regroup if necessary.

1. 14
    + 37
    _____

2. 29
    + 17
    _____

3. 25
    + 73
    _____

4. 46
    + 27
    _____

5. 81
    + 18
    _____

Regroup the ones to show more tens.

6. 6 tens 12 ones = _____ tens _____ ones

7. 8 tens 16 ones = _____ tens _____ ones

THINK
10 ones = 1 ten

Solve. Remember to label your answers.

8. In Kay's class there are 16 girls and 14 boys.
    How many students are in her class?

    _____

9. Kendra rode her bicycle for 26 minutes before school.
    After school, she rode her bicycle for 45 minutes. How
    many minutes did she ride her bicycle in all?

    _____

10. Jacob was given a $50 bill to buy clothes. He picked out a
    shirt for $16 and a pair of pants for $38. Does Jacob have
    enough money to pay for the clothes? Explain.

    _____

    _____

Name _____

Adding by Regrouping Tens

Add. Regroup if necessary.

1. 84	2. 51	3. 31	4. 53	5. 82
+ 33	+ 56	+ 68	+ 75	+ 15

Regroup the tens as hundreds and tens.

6. 15 tens = _____ hundreds _____ tens   THINK
                                                   10 tens = 1 hundred

7. 33 tens = _____ hundreds _____ tens

Solve. Remember to label your answers.

8. Susan delivered 83 newspapers and Michael delivered 52 newspapers. How many newspapers did they deliver altogether?

_____

9. Students at Smithville School gathered cans for a food drive. The third-graders brought in 93 cans and the fourth-graders brought in 92 cans. How many cans did the third- and fourth-graders bring in altogether?

_____

10. The bears at the zoo ate 86 pounds of food one day and 82 pounds the next day. How many pounds of food did the bears eat during the two days?

_____

Name _____

Adding with Two Regroupings

Add.

1.
$$
\begin{array}{r}
58 \\
+ 64 \\
\hline
\end{array}
$$

2.
$$
\begin{array}{r}
84 \\
+ 16 \\
\hline
\end{array}
$$

3.
$$
\begin{array}{r}
27 \\
+ 94 \\
\hline
\end{array}
$$

4.
$$
\begin{array}{r}
476 \\
+ 249 \\
\hline
\end{array}
$$

5. $165 + 279 =$ _____

6. $194 + 587 =$ _____

7. $368 + 279 =$ _____

Solve. Remember to label your answers.

8. Nick has 196 baseball cards in his collection. His sister has 199 cards. How many cards do they have altogether?

_____

9. In the morning, 427 people entered the aquarium. In the afternoon, 496 people entered. How many people went to the aquarium that day?

_____

MIXED Practice

Write the place value of the 7 in each number.

1. 6,703

_____

2. 4,671

_____

3. 7,149

_____

4. 789,123

_____

5. 10,247

_____

6. 27,192

_____

Name _____

Adding with Three or More Addends

Add. Look for sums of 10.

1.	18	2.	34	3.	162	4.	702	5.	656
	22		17		31		334		123
	11		22		355		565		431
	+ 20		+ 44		+ 209		+ 111		+ 202

6.	274	7.	305	8.	722	9.	64	10.	333
	186		624		908		261		471
	+ 451		+ 875		+ 453		+ 897		+ 226

Use the chart to solve. Remember to label your answers.

Carnival Attendance	
Day	**Number of People**
Wednesday	245
Thursday	316
Friday	596
Saturday	799

11. What is the total number of people who attended the carnival on Thursday, Friday, and Saturday?

_____

12. How many more people attended on Friday than Wednesday?

_____

13. What was the total attendance for all four days of the carnival?

_____

14. Did more people attend the carnival on Wednesday and Thursday or on Friday and Saturday?

_____

Name _____

Ladybug's Path

Help Ladybug add her way to 2,000. She needs to find a path through the bush to get her to the end of the leaf maze. She can only move up, down, left, or right, and can only move one square at a time. She cannot move diagonally. Can you get her where she needs to go?

0	541	422	33	150
Start here ↓ 102	299	264	111	130
148	125	361	90	35
314	17	258	43	230

2,000 End

Hint: Ladybug can do it in 11 moves. Keep track of your addition below. Notice that Ladybug hits exactly 1,000 on her fifth move.

1. 0 + <u>102</u> = ____

2. <u>102</u> + ____ = ____

3. ____ + ____ = ____

4. ____ + <u>361</u> = ____

5. ____ + ____ = 1,000

6. ____ + ____ = ____

7. ____ + ____ = <u>1,455</u>

8. ____ + ____ = ____

9. ____ + <u>130</u> = ____

10. ____ + ____ = ____

11. <u>1,770</u> + ____ = 2,000   **YOU DID IT!**

Name _____

Adding Greater Numbers

Add.

1. 4,218
   + 3,701

2. 8,002
   + 1,979

3. 3,370
   + 6,439

4. 75,567
   + 4,134

5. 43,724
   + 42,397

6. 43,126
   + 5,878

7. 1,212
   + 878

8. 65,979
   + 12,466

Solve. Remember to label your answers.

9. One year the Smiths traveled 2,468 miles on their vacation. The next year they traveled 3,802 miles. How far did they travel over the two years?

10. The local newspaper prints 24,625 papers on Saturday and 45,316 on Sunday. How many papers are printed in a weekend?

11. On Sunday, Clint scored 18,251 points on his video game. On Friday, he scored 18,989 points. What was the combined score?

12. The grocery store clerks shelved 3,599 items on Tuesday and 6,499 items on Wednesday. How many items did they shelve over the two days?

MIXED Practice

Estimate each sum by rounding.

1. 76
   + 33 + _____

2. 228
   + 116 + _____

3. 53
   + 48 + _____

Name _____

Problem-Solving Skill: Is the Answer Reasonable?

Ring the letter of the most reasonable answer.

1. A baker made 360 glazed donuts and 480 chocolate donuts. How many donuts did he make in all?

   a. 840 donuts   b. 700 donuts   c. 84 donuts

2. Sam sold 125 chocolate cones, 179 vanilla cones, and 215 strawberry cones at the ice cream shop. How many cones did Sam sell in all?

   a. 619 cones   b. 519 cones   c. 419 cones

3. Mr. Moore counts the apples he picks each day from his orchard. This week he counted these amounts: 110, 80, 90, 120, and 130. How many apples did Mr. Moore pick this week?

   a. 430 apples   b. 530 apples   c. 630 apples

4. Alex displays his seashell collection in 2 cases. He has 82 shells in one case and 105 shells in the other. How many shells does Alex have in his collection?

   a. 287 shells   b. 187 shells   c. 87 shells

5. Amy has 12 belts, 6 purses, 4 dresses, 8 skirts, and 10 shirts in her closet. How many items does Amy have in her closet altogether?

   a. 35 items   b. 40 items   c. 100 items

Name _____

Chapter 2 Review

Add.

1.	3	**2.**	24	**3.**	24	**4.**	4	**5.**	72
	8		+ 67		39		+ 4		68
	+ 6				+ 52				+ 45

6.	9	**7.**	7	**8.**	9	**9.**	8	**10.**	4
	3		+ 7		+ 8		+ 3		+ 5
	+ 8								

11. $8 + (4 + 6) =$ _____

12. $(4 + 1) + 8 =$ _____

Complete each fact family.

13. $8 + 3 = 11$ _____

14. $17 - 9 = 8$ _____

_____ _____

_____ _____

Compare. Write $<$, $>$, or $=$.

15. $9 + 3 \bigcirc 8 + 5$

16. $4 + 7 \bigcirc 8 + 2$

17. $6 + 6 \bigcirc 7 + 5$

18. $3 + 4 \bigcirc 2 + 6$

Add. Regroup if necessary.

19.	25	**20.**	537	**21.**	46,015	**22.**	86
	+ 86		+ 21		+ 3,291		+ 75

Add.

23.   13
    + 70

24.   2,650
    +   48

25.   48
    + 2

Add.

26.   73
    + 54

27.   4,568
    + 730

28.   685
    + 217

29.   54,197
    + 3,066

Estimate each sum by rounding.

30.   292
    + 817   + _____

31.   633
    + 288   + _____

Ring the letter of the most reasonable answer.

32. Debbie's family drove 250 miles before lunch. After lunch they drove another 375 miles. How far did they drive that day?

    a. 475 miles   b. 625 miles   c. 900 miles

Solve. Remember to label your answers.

33. Henry and Sarah were buying balloons for a birthday party. Henry bought 6 blue balloons and 9 purple balloons. Sarah bought 5 yellow balloons and 8 orange balloons. Who bought more balloons?

    _____

34. In a train yard, there are three engines with 6 wheels each and two engines with 10 wheels each. How many wheels are there in the train yard in all?

    _____

Name _____

Chapter 2 Test Prep

Ring the letter of the correct answer.

1. Name the property that is shown. $24 + 8 = 8 + 24$

    a. Identity Property of Addition

    b. Associative Property of Addition

    c. Commutative Property of Addition

    d. none of these

2. Which math fact can help you solve $16 + 7$?

    a. $7 + 7$ b. $6 + 7$

    c. $10 + 6$ d. $17 + 1$

3. $24 + 67 = $ _____

    a. 81   b. 83

    c. 91   d. 93

4. $239 + 68 = $ _____

    a. 337   b. 317

    c. 307   d. 297

5. 6 + (7 + 5) = _____

   a. 15   b. 16

   c. 17   d. 18

6. 4,568 + 730 = _____

   a. 5,398   b. 5,298

   c. 4,298   d. 4,238

7. Estimate. 29 + 83 = _____

   a. 120   b. 110

   c. 100   d. 90

8. Estimate. 81,408 + 18,673 = _____

   a. 100,000   b. 90,000

   c. 80,000   d. 20,000

9. Which is a member of the fact family containing 8 + 5 = 13?

   a. 8 − 5 = 3   b. 13 − 8 = 5

   c. 13 + 5 = 18   d. 18 + 5 = 23

10. Maria made wooden flowers at the craft fair. She made
    38 flowers on Monday, 26 on Tuesday, 41 on Wednesday,
    and 35 on Thursday. Which answer is the most reasonable
    for the number of flowers she made in all?

    a. 200 flowers   b. 140 flowers

    c. 120 flowers   d. 100 flowers

Name _____

Subtraction Strategies

Subtraction Strategies	
Strategy	**Example**
a. **Count back** • Start with the greater number and count back. • Use for subtracting 1, 2, 3, or 4.	15 − 4 = ___ 15: 14, 13, 12, 11 15 − 4 = 11
b. **Count up** • Start with the lesser number and count up to the greater number. • Use when the numbers are close to each other.	14 − 12 = ___ 12: 13, 14 14 − 12= 2
c. **Related fact** • Use a related fact from a fact family.	13 − 7 = ___ 7 + 6 = 13 13 − 7 = 6

Subtract. Write the letter of the subtraction strategy you used.

____ 1. 11 − 3 = ____   ____ 2. 15 − 12 = ____

____ 3. 37 − 2 = ____   ____ 4. 45 − 3 = ____

____ 5. 18 − 15 = ____   ____ 6. 11 − 2 = ____

____ 7. 8 + 8 = 16, so 16 − 8 = ____

____ 8. 6 + 7 = 13, so 13 − 6 = ____

Subtract.

9. 19 − 1 = ____   10. ____ = 26 − 23   11. 17 − 8 = ____

12. ____ = 53 − 3   13. 18 − 4 = ____   14. ____ = 11 − 5

15. 39 − 35 = ____   16. ____ = 75 − 2   17. 13 − 11 = ____

Name _____

More Subtraction Strategies

More Subtraction Strategies	
Strategy	**Example**
a. Easy 9s • When subtracting 9 from a number that is between 10 and 18, add the digits of the two digit number.	Add the digits in 15. $15 - 9 =$ ___ $1 + 5 = 6$ $15 - 9 = 6$
b. Subtracting 9 from larger numbers • When subtracting 9, first subtract 10 and then add 1.	Subtract 10. $35 - 9 =$ ___ $35 - 10 = 25$ Add 1. $25 + 1 = 26$ $35 - 9 = 26$
c. Patterns • To subtract larger numbers, first think of a simpler subtraction fact and find a pattern.	$64 - 10 = 54$ $8 - 5 = 3$ $64 - 20 = 44$ $28 - 5 = 23$ so, $64 - 30 = 34$ so, $128 - 5 = 123$ Pattern: keep the ones, subtract the tens. Pattern: subtract the ones, keep the tens and hundreds.

Subtract. Write the letter of the subtraction strategy you used.

___ **1.** $16 - 9 =$ ___   ___ **2.** $85 - 40 =$ ___

___ **3.** $53 - 9 =$ ___   ___ **4.** $78 - 9 =$ ___

Subtract. Use a pattern.

5. $7 - 3 =$ ___ $17 - 3 =$ ___ $27 - 3 =$ ___ $37 - 3 =$ ___

What pattern did you use? _____

6. $15 - 8 =$ ___ $115 - 8 =$ ___ $215 - 8 =$ ___ $315 - 8 =$ ___

What pattern did you use? _____

Subtract.

7. 56 − 20 = _____   8. _____ = 34 − 9

9. 48 − 9 = _____   10. _____ = 13 − 9

11. 75	12. 98	13. 63	14. 87	15. 67
− 40	− 60	− 50	− 30	− 20

16. When you subtract a number with zero ones, what do you notice about the difference?

    _____

Solve. Remember to label your answers.

17. Kim subtracted 40 from 99. She said the difference was 50.
    Is Kim correct? Explain your answer.

    _____

18. There are 78 children from Calvert School visiting the aquarium.
    Nine of them are looking at the shark exhibit. How many of them are
    not looking at the shark exhibit?

    _____

19. Bruno weighs less than Spot. The sum of their weights is 220 lbs.
    The difference in their weights is 10 pounds. How much does each
    dog weigh?

    Bruno weighs _____.   Spot weighs _____.

20. There are 57 penguins at the zoo. There are 37 penguins swimming
    in the water. How many penguins are *not* in the water?

Name _____

Mental Subtraction

You can use mental math to find the difference between two amounts that are close to the same number.

Example: 58 − 45 = _____

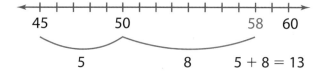

THINK   From 45 to 50 is 5.
From 50 to 58 is 8.
5 + 8 = 13
58 − 45 = 13

What is 120 − 95?

From 95 to 100 is 5.
From 100 to 120 is 20.
5 + 20 = 25
120 − 95 = 25

> You can use a "stepping stone" to make the problem easier. What number, or stepping stone, between 95 and 120 can you use to solve the problem?

Find each difference.

1. 89 − 74 = _____

2. 63 − 58 = _____

3. _____ = 35 − 18

4. _____ = 76 − 51

5. 93 − 76 = _____

6. 41 − 27 = _____

7. _____ = 130 − 94

8. _____ = 150 − 135

Solve.

9. What number would be a good stepping stone

to help you find 1,245 − 995? _____

10. 1,245 − 995 = _____

Name _____

Estimating Differences by Rounding

Estimate each difference by rounding to the greatest shared place value. Ring the best estimate.

1.   69
       − 27

    a. 70
    b. 40
    c. 30

2.   738
        − 21

    a. 700
    b. 710
    c. 720

3.   400
        − 78

    a. 300
    b. 330
    c. 320

4.   891
       − 427

    a. 500
    b. 400
    c. 300

Estimate each difference by rounding to the greatest shared place value. Use your estimate to select the exact difference.

5.   75
       − 48

    a. 37
    b. 47
    c. 27

6.   891
       − 572

    a. 619
    b. 319
    c. 719

7.   333
       − 111

    a. 322
    b. 122
    c. 222

8.   680
       − 59

    a. 621
    b. 521
    c. 421

Name _____

Subtracting by Regrouping Tens

First, subtract the digits in the ones place. If the digit on top (minuend) is less than the digit on the bottom (subtrahend), regroup, and then subtract.

$$\begin{array}{r} 47 \\ -19 \\ \hline \end{array}$$ Regroup? 7 < 9 Yes.

Now subtract.
$$\begin{array}{r} {}^{3\ 17} \\ \cancel{47} \\ -19 \\ \hline 28 \end{array}$$

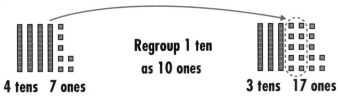

4 tens 7 ones   3 tens 17 ones

You can regroup 47 as 3 tens 17 ones.
Then you can subtract 17 − 9.

Regroup 1 ten as 10 ones.

Ex:

Tens	Ones		Tens	Ones
2	3	=	1	13

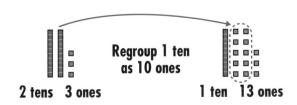

2 tens 3 ones   1 ten 13 ones

Regroup. Write the missing numbers.

1.

Tens	Ones		Tens	Ones
5	6	=	4	

2.

Tens	Ones		Tens	Ones
3	8	=		18

3.

Tens	Ones		Tens	Ones
6	2	=		

4.

Tens	Ones		Tens	Ones
4	5	=		

5.

Tens	Ones		Tens	Ones
7	1	=		

6.

Tens	Ones		Tens	Ones
8	4	=		

Subtract. Regroup if necessary.

7.   64
    − 27

8.   38
    − 17

9.   76
    − 49

10.   57
    − 17

11.   60
    − 28

12.   84
    − 26

13.   92
    − 64

14.   39
    − 18

15.   40
    − 24

16.   85
    − 49

17. 54 − 31 = _____ −_____

18. 85 − 17 = _____ −_____

MIXED Practice

Solve. Remember to label your answers.

1. Fred sells snowballs at the zoo. Today he sold 14 vanilla, 39 strawberry, and 46 chocolate snowballs. How many snowballs did he sell today?

    _____

2. On Monday, Beau picked 157 melons in the morning and 10 more in the afternoon. How many melons did he pick on Monday?

    _____

3. Jason mowed 4 lawns and earned $20, $25, $20, and $30. How much money did he earn altogether?

    _____

Name _____

Subtracting 3-Digit Numbers

Subtract. Regroup if necessary.

1.	493 − 367	2.	370 − 124	3.	936 − 409	4.	$724 − 203	5.	$713 − 309

6.	940 − 615	7.	247 − 129	8.	483 − 150	9.	340 − 125	10.	952 − 716

Solve. Remember to label your answers.

11. Vince's team scored 151 runs this season. Last season they scored 98. How many more runs did they score this season?

_____

12. On Monday, Al's T-Shirt Shop received a shipment of 325 shirts. On Wednesday, they received a shipment of 175 shirts. How many more shirts were received on Monday than on Wednesday?

_____

MIXED Practice

Skip-count to fill in the missing numbers.

1. 245, 250, 255, _____, _____

2. 300, _____, 350, 375, _____

3. 890, 910, _____, _____, 970

4. 435, _____, 505, 540, _____, 610

Name _____

Problem-Solving Skill: When to Estimate

Decide if you need an exact answer or an estimated answer.
Write *exact* or *estimate*. Then solve.

1. Carter is saving money to buy a new bike. He has $87 saved. The bike costs $165. How much more money does Carter need to buy the bike?

   _____

   _____

2. A printing company printed 3,000 programs to sell at a baseball game. They sold 2,147 programs. About how many programs were left over?

   _____

   _____

3. The Edwards family drove 176 miles on Friday. Then they drove 89 miles on Saturday. About how many more miles did they drive on Friday than Saturday?

   _____

   _____

4. A carpenter cuts a 96-inch piece of wood into two pieces. One piece measures 38 inches. How long is the other piece of wood?

   _____

   _____

5. Riley rented two movies, a comedy and a drama. The comedy is 115 minutes long and the drama is 94 minutes long. How much longer is the comedy?

   _____

   _____

6. An aquarium had 8,952 visitors yesterday. Today, the number of visitors was 1,356 less than yesterday. About how many people visited the aquarium today?

   _____

   _____

© Calvert School

Name _____

Subtracting by Regrouping Tens and Hundreds

It is not possible to subtract 6 tens from 2 tens.
So, regroup 4 hundreds 2 tens as 3 hundreds 12 tens.

$$
\begin{array}{r} 4\,2\,6 \\ -\,1\,6\,3 \\ \hline 3 \end{array}
\qquad
\begin{array}{r} {\scriptstyle 3\,12} \\ \cancel{4}\cancel{2}\,6 \\ -\,1\,6\,3 \\ \hline 2\,6\,3 \end{array}
$$

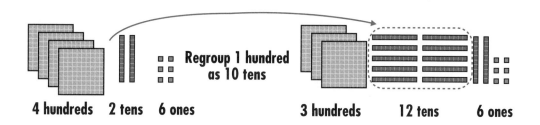

4 hundreds   2 tens   6 ones   **Regroup 1 hundred as 10 tens**   3 hundreds   12 tens   6 ones

Regroup 1 hundred as 10 tens. Write the missing numbers.

Example:

Hundreds	Tens	Ones		Hundreds	Tens	Ones
7	4	3	=	6	14	3

1.

Hundreds	Tens	Ones		Hundreds	Tens	Ones
6	3	7	=			7

2.

Hundreds	Tens	Ones		Hundreds	Tens	Ones
8	3	6	=			6

3.

Hundreds	Tens	Ones		Hundreds	Tens	Ones
9	8	2	=			2

4.

Hundreds	Tens	Ones		Hundreds	Tens	Ones
5	6	9	=			9

© Calvert School

Subtract. Regroup if necessary.

5. 946
    − 381

6. 611
    − 457

7. 821
    − 475

8. 762
    − 595

9. 868
    − 568

10. 630
    − 470

11. 586
    − 342

12. 322
    − 196

13. 623
    − 459

14. 892
    − 678

15. 800
    − 50

16. 740
    − 460

Solve. Remember to label your answers.

17. A pet store had 475 boxes of fish food on the shelf.
    At the end of the week, 293 boxes were still on the
    shelf. How many boxes of fish food were sold
    during the week?

    _____

18. Manny counts the bags of dog food at his pet store.
    He counts 352 bags. Then 225 more bags are delivered
    to his store. How many bags of dog food are in Manny's
    store now?

    _____

Name _____

Missing Digits

Write the missing numbers.

1.
```
    3 9
  + 2 □
  -----
    6 2
```

2.
```
    7 6
  + □ 1
  -----
    9 7
```

3.
```
    5 □
  + 3 8
  -----
    9 3
```

4.
```
    3 □
  + □ 4
  -----
    8 2
```

5.
```
    2 7
  + 3 □
  -----
    6 3
```

6.
```
    □ 1
  + 5 9
  -----
  1 4 0
```

7.
```
    8 3
  - 2 □
  -----
    5 7
```

8.
```
    9 0
  - □ 4
  -----
    6 □
```

9.
```
    3 □ 7
  - 2 8 4
  -------
    1 1 3
```

10.
```
    □ 4 5
  - 1 9 7
  -------
    3 4 □
```

11.
```
    4 □ 8
  - 2 2 5
  -------
    2 6 □
```

12.
```
    9 4 □
  - □ 6 7
  -------
    6 8 2
```

13.
```
    5 □ 6
  - 1 7 3
  -------
    3 5 □
```

14.
```
    □ 1 4
  - 2 2 □
  -------
    1 8 9
```

15.
```
    9 □ 7
  - 4 3 □
  -------
    5 3 2
```

16.
```
    7 7 2
  -   □ 1
  -------
    6 9 1
```

Name _____

Subtracting Across Zeros

Sometimes it is necessary to regroup hundreds to subtract ones.
First regroup 1 hundred as 10 tens.

$$\begin{array}{r} 405 \\ -186 \\ \hline \end{array}$$

Then regroup 1 ten as ten ones. Now you can subtract.

Step 1   Step 2

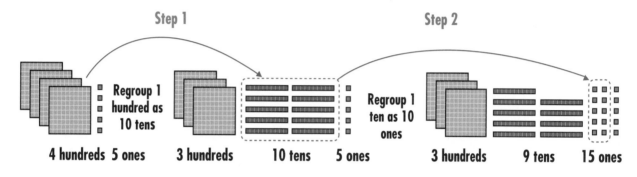

4 hundreds 5 ones   3 hundreds   10 tens   5 ones   3 hundreds   9 tens   15 ones

The subtraction problem looks like this:

$$\begin{array}{r} \overset{9}{} \\ 3\ \ \cancel{10}\ 15 \\ \cancel{4}\ \cancel{0}\ \cancel{5} \\ -1\ 8\ 6 \\ \hline 2\ 1\ 9 \end{array}$$

Regroup 1 hundred as 10 tens. Then regroup 1 ten as 10 ones.

	Step 1	Step 2
	5 10	9
		5 10 12
Ex.: 602	$\cancel{6}\ \cancel{0}\ 2$	$\cancel{6}\ 0\ \cancel{2}$

1. 203

2. 607

3. 804

Subtract. Regroup if necessary.

4.   302
      − 75

5. $806
      − 328

6.   504
      − 178

7.   201
      − 135

8.   601
      − 273

9.   807
      − 529

10.   202
       − 125

11.   507
       − 246

12.   905
       − 68

13.   501
       − 276

14. $109
       −   69

15.   307
       −   49

16. $802
       − 363

17.   701
       − 403

18.   600
       − 300

19.   405
       − 67

Solve. Remember to label your answer.

20. Mr. Anderson wants to collect 600 pounds of newspaper to recycle. So far, he has collected 314 pounds. How many more pounds of newspaper does he need to collect to reach his goal?

21. On Saturday, 705 cars paid the toll to go through a tunnel. On Sunday, 639 cars paid the toll. How many more cars paid the toll on Saturday than on Sunday?

Name _____

Subtracting Greater Numbers

Subtract.

1.   8,452
    − 3,815
	2. 6,842
	    − 3,674
		3. 7,526
		    − 3,238
			4. 5,437
			    − 2,169

5.   86,719
    − 24,305
	6. 5,888
	    − 3,269
		7. 53,904
		    − 41,237
			8. 8,523
			    − 7,845

9.   8,204 − 3,427 = _____

10.   8,508 − 6,294 = _____

Solve. Remember to label your answers.

11. Scenic Highway is 3,250 miles long. Forest Highway is 2,647 miles long. How much longer is Scenic Highway than Forest Highway?

12. The park rangers caught two bears. One bear weighed 347 pounds and the other weighed 500 pounds. What is the difference in the bears' weights?

Name _____

Checking Subtraction with Addition

Addition and subtraction are opposites. Subtraction takes numbers apart and addition puts numbers together. Addition can be used to check the answer to a subtraction problem.

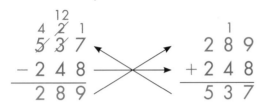

```
      12
   4  2  1
   5̶  3̶  7          2  8  9
 − 2  4  8        + 2  4  8
   2  8  9          5  3  7
```

Add the difference to the amount that was subtracted. The sum should be the same as the top number in the original subtraction problem. If not, subtract again to find the mistake.

More Examples

A.
```
        3 11
     8  4̶  1̶           1
                    2  1  7
   − 6  2  4        + 6  2  4
     2  1  7          8  4  1
```

B.
```
          9
     5  1̶0̶ 11
     6̶  0̶  1̶           1  1
                    2  3  6
   − 3  6  5        + 3  6  5
     2  3  6          6  0  1
```

Subtract. Check with addition.

		Check			Check			Check
1.	743		**2.**	619		**3.**	854	
	− 427 +			− 425 +			− 395 +	

		Check			Check			Check
4.	7,510		**5.**	7,709		**6.**	12,845	
	− 3,198 +			− 4,463 +			− 8,486 +	

Name _____

Problem-Solving Skill: Identifying Missing Information

Solve and check using the opposite operation. If the problem cannot be solved, identify the missing information.

1. Happy Pets pet store has 126 goldfish and 96 guppies. Smiling Pets pet store has 58 goldfish. How many more guppies does Happy Pets have than Smiling Pets?

2. John donated 1,028 pennies to the animal shelter. Emily donated 1,268 pennies. How many more pennies did Emily donate than John?

3. Herb, Marsha, and Simon have baseball cards. Herb has 124 cards and Marsha has 96 cards. How many more cards does Herb have than Simon?

4. On Wednesday, Ryan moved 175 pieces of wood. On Thursday, he moved some more pieces of wood. How many total pieces of wood did Ryan move?

5. A cookie recipe calls for 2 cups of sugar. Mr. Land wants to make a double batch of cookies. How much sugar does he need?

6. Gayle's dentist appointment is at 10:00 A.M. The dentist will take $\frac{1}{2}$ hour to examine Gayle's teeth. What time does Gayle need to leave home to get to her appointment on time?

Name _____

Problem-Solving Skill: Does the Answer Make Sense?

Circle the answer that makes the most sense.

1. How many children are on Coach McHenry's softball team?

   a. 15   b. 150   c. 1,500

2. Marie went fishing in a river near her house. What was the weight of one of the fish she caught?

   a. 3 pounds   b. 30 pounds   c. 300 pounds

3. Artie rode his bike through his neighborhood for about an hour. About how far did he ride?

   a. 400 miles   b. 40 miles   c. 4 miles

4. How many people fit in a stadium where professional baseball is played?

   a. 300   b. 1,000   c. 45,000

5. Denny's family likes to watch TV. How many television sets do they have in their house?

   a. 4   b. 20   c. 50

6. Martin had strawberries for lunch. How many strawberries did Martin eat?

   a. 8 strawberries   b. 80 strawberries   c. 800 strawberries

Name _____

Chapter 3 Review

1. Which of these are subtraction problems?

   a. Jim and Hank had 37 gum balls. They chewed 18. How many are left?

   b. Jim has 55 gum balls and Hank has 45 sticks of gum. How many pieces of gum do they have in all?

   c. Jim and Hank have 95 CDs. Thirty-seven of them are Hank's. How many belong to Jim?

   d. Jim has 423 baseball cards and Hank has 615. How many more cards does Hank have than Jim?

Subtract.

2. $57 - 30 =$ _____   3. $89 - 40 =$ _____   4. $90 - 56 =$ _____

Subtract. Regroup if necessary.

5.	6.	7.	8.
55 − 27	82 − 73	851 − 732	$63 − 48

9.	10.	11.	12.
93 − 75	6,473 − 3,245	$551 − 343	8,635 − 5,264

13.	14.	15.	16.
7,318 − 1,624	8,516 − 1,273	19,449 − 2,772	7,349 − 5,155

Estimate each difference by rounding to the greatest shared place value.

17.	$6,567 − 372 − _____	18.	45 − 28 − _____	19.	540 − 91 − _____

20.	981 − 205 − _____	21.	624 − 38 − _____	22.	7,291 − 5,981 − _____

Solve. Remember to label your answer.

23. Nancy had 150 tickets to sell. She sold 79. How many
   more tickets does she have to sell?

   _____

**Decide if you need an exact answer or an estimated answer. Write *exact*
or *estimate*. Then solve.**

24. Harold drove 1,315 miles during his vacation last year and
   2,109 miles during his vacation this year. About how many
   more miles did he drive during his vacation this year?

   _____

25. An auditorium has 4,000 seats. There were 871 empty seats at
   a concert. How many seats were filled at the concert?

   _____

Name _____

Chapter 3 Test Prep

Ring the letter of the correct answer.

1. $\begin{array}{r} 23 \\ -\ 10 \\ \hline \end{array}$ **a.** 13 **b.** 17 **c.** 20 **d.** 33

2. $\begin{array}{r} 67 \\ -\ 30 \\ \hline \end{array}$ **a.** 23 **b.** 34 **c.** 37 **d.** 40

3. Estimate the difference by rounding.
   5,967 – 1,134 = _____

   a. 6,000 **b.** 5,000

   c. 4,000 **d.** 3,000

4. $\begin{array}{r} 4{,}579 \\ -\ 2{,}887 \\ \hline \end{array}$

   a. 1,692 **b.** 1,792

   c. 2,312 **d.** 2,792

5. On Mr. Tyding's farm are 845 chickens. How many chickens are left if he sells 643?

   a. 199 chickens

   b. 202 chickens

   c. 302 chickens

   d. 342 chickens

6. Mickie sells vanilla and chocolate ice cream cones. On Tuesday, she sold 189 cones; 34 were vanilla. How many cones were chocolate?

   a. 155 cones **b.** 163 cones

   c. 165 cones **d.** 175 cones

7. $\begin{array}{r} 498 \\ -\ 159 \\ \hline \end{array}$

   a. 349 **b.** 339

   c. 312 **d.** 302

8. $\begin{array}{r} 3{,}246 \\ -\ 2{,}769 \\ \hline \end{array}$

   a. 534 **b.** 477

   c. 377 **d.** 368

9. A third-grade class is selling pencils. They started with 288 pencils. How many pencils are left to sell? Which missing fact is needed to solve the problem?

   a. the number of pencils in each box

   b. the number of students in third grade

   c. the number of pencils sold so far

   d. Nothing is missing. The problem can be solved.

10. Adrienne needs to make 50 party favors. She has made 18 favors so far. About how many more favors does she need to make?

    a. about 10 more favors   b. about 20 more favors

    c. about 30 more favors   d. about 40 more favors

Name _____

Counting Money

Write the amount of money.

1.

   _____ ¢ $ _____ . _____

2.

   _____ ¢ $ _____ . _____

3.

   $ _____ . _____

4.

   $ _____ . _____

5.

   $ _____ . _____

6.

   $ _____ . _____

7. 2 twenty-dollar bills, 4 one-dollar bills, 1 nickel, 2 pennies

   $ _____ . _____

8. 3 ten-dollar bills, 2 five-dollar bills, 1 dime, 1 penny

   $ _____ . _____

Write how many dollars, dimes, and pennies.

9. 32¢ = _____ dimes _____ pennies

10. 44¢ = _____ dimes _____ pennies

11. $7.29 = _____ dollars _____ dimes _____ pennies

12. $5.06 = _____ dollars _____ dimes _____ pennies

Read and write each amount.

	Dollars	Dimes	Pennies	
13.	4	5	6	$. _____

	Dollars	Dimes	Pennies	
14.	3	0	8	$. _____

	Dollars	Dimes	Pennies	
15.	7	2	0	$. _____

	Dollars	Dimes	Pennies	
16.	8	2	8	$. _____

17. fifty-three cents $. _____

18. eighty-nine cents $. _____

19. five dollars and eighty-four cents $. _____

20. six dollars and twenty-five cents $. _____

Solve. Remember to label your answers.

21. Erin has 120 pennies, 2 nickels, and 1 dime in her bank. How much money does Erin have?

22. Dot has 75¢. She has 2 quarters and some nickels. How many nickels does Dot have?

MIXED Practice

Add or subtract.

1. 328
  + 247

2. 500
  − 382

3. 532
  − 75

Compare. Write <, >, or = .

4. 634 ◯ 643

5. 1,072 ◯ 1,172

Name _____

Comparing Money Amounts

Write the letter of the group that shows an equal amount.

_____ 1.     a.

_____ 2.     b.

_____ 3.    c.

Compare the amounts of money using <, >, or =.

4.  _____

5.  _____

Draw bills and coins to show the amounts in two ways.

		Way 1	Way 2
6.	$1.37		
7.	$2.96		
8.	$0.53		
9.	$0.79		

Name _____

Using Coins

Follow the directions in each column. Use your punch-out coins or real money to show each amount. Draw each coin by making a circle with the amount written inside.

(1¢) (5¢) (10¢) (25¢) (50¢)	Remember, there is a coin called a *half dollar* that is worth 50¢
Column A: First do this... Use the *least* number of coins to make the amount shown.	**Column B: Then do this...** Use the specific number of coins indicated at the top of each row to make the amount shown.

1. 31¢		using 6 coins
2. 71¢		using 7 coins
3. 94¢		using 10 coins
4. 22¢		using 14 coins
5. 55¢		using 4 coins

Name _____

Making Change

Find the amount of change you would receive for each purchase.
Use play money to help you. Use as few coins as possible.

Cost	Money Given	Amount of Change

1.       _____

2.       _____

3.       _____

4. $3.20      _____

5. $1.37      _____

6. five dollars and seventy-five cents      _____

7. nine dollars and eleven cents   nine dollars and fifty cents   _____

8.       _____

Name _____

Count Your Change

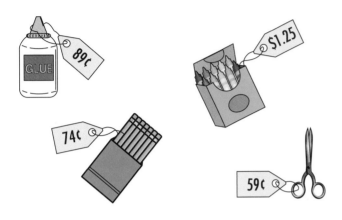

Find the amount of change.

1. You buy: 1 pack of pencils
   You give the clerk: $1.00

   change: _____

2. You buy: 1 ruler
   You give the clerk: 50¢

   change: _____

3. You buy: 1 bottle of glue
            1 notebook
   You give the clerk: $5.00

   change: _____

4. You buy: 1 pack of crayons
            1 ruler
            1 pair of scissors
   You give the clerk: $3.00

   change: _____

5. You buy: 1 notebook
            1 box of crayons
   You give the clerk: $5.00

   change: _____

6. You buy: 1 bottle of glue
            2 packs of crayons
   You give the clerk: $5.00

   change: _____

Name _____

Adding and Subtracting Money

Add or subtract.

1. $0.78
   + 2.32

2. $2.19
   + 0.51

3. $4.00
   + 2.80

4. $2.33
   + 3.71

5. $2.00
   − 0.63

6. $23.46
   − 0.25

7. $42.39
   − 21.28

8. $13.37
   − 10.40

9. $64.15
   − 31.96

10. $213.79
    + 326.48

11. $2.63 + $3.15 = _____

12. $3.10 − $1.95 = _____

Solve. Remember to label your answers.

13. Kevin wants to buy a DVD that costs $9.90. So far he has saved $6.57. How much more does he need?

14. Chuck took 2 quarters, 7 dimes, 9 nickels, and 22 pennies out of his bank. How much money did he take out?

MIXED Practice

Solve.

1. Write the number that has 7 in the hundreds place, 3 in the tens place, 1 in the thousands place, and 2 in the ones place.

2. Lila is the eighth person in line. You are in front of Lila. What is your position in the line?

Name _____

Problem-Solving Application: Earning and Spending Money

Solve. Remember to label your answers.

1. Jack mowed 5 lawns one week and 7 lawns the following week. He charged $10.00 per lawn. How much money did he make each week? How much did he make altogether?

   _____

2. Ted had $20.00 in his bank. He bought lunch for his mother on Mother's Day. He had $5.72 left after lunch. How much did he spend on lunch?

   _____

3. Jill wants to buy a pair of shoes that cost $19.99. So far, she has saved 2 five-dollar bills, 2 one-dollar bills, 3 dimes, 2 nickels, and 3 pennies. How much has she saved? How much more does she need to buy the shoes?

   _____

4. Denise had 3 ten-dollar bills, 2 five-dollar bills, 3 one-dollar bills, 2 quarters, 2 dimes, 3 nickels, and 4 pennies.

   a. How much money did she have? _____

   b. She spent $15.65 to buy groceries and $13.50 to go to the movies. How much money did she spend altogether? _____

   c. How much money does Denise have left? _____

5. On Saturday, Jane babysat and earned $12.40. Sunday afternoon she bought a book for $6.75. Monday through Friday she fed her aunt's cat and earned $2.50 per day. How much money did Jane have at the end of the week?

   _____

Name _____

Telling Time to Five Minutes

Ring the letter of the correct answer.

1. The hour hand is the ____.   **a.** longer hand   **b.** shorter hand

2. The minute hand is the ____.   **a.** longer hand   **b.** shorter hand

Write the letter of the matching time.

3. ten forty-five ____   **a.** 9:15

4. nine fifteen ____   **b.** 12:30

5. twelve thirty ____   **c.** 10:45

6. twelve forty-five ____   **d.** 12:45

Write the time on the digital clock.

Reminder: There are 5 minutes between each number on a clock.

7.

8.

9.

10.

11.

12.

Draw the minute hand to show the time on each digital clock.

13. `7:30`

14. `4:15`

15. `10:35`

16. `8:50`

17. `11:20`

18. `5:05`

Write the missing times to complete the pattern.

19. 6:20, 6:30, 6:40, _____, _____

20. 12:45, 12:50, 12:55, _____, _____

Name _____

Telling Time to the Minute

Reminder: When telling time to the minute, remember to start at 12 and move to the right until you reach the minute hand. Count by fives for each number on the clock. If the minute hand is between two numbers, count on one minute for each little line. Stop when you get to the minute hand.

Write the time.

1.

___:___

2.

___:___

3.

Circle the letter of the correct time.

4.

a. 10:18

b. 3:52

5.

a. 7:19

b. 3:35

6.

a. 6:57

b. 11:32

Draw the hour and minute hand to show the given time.

7.

10:43

8.

2:01

9.

5:28

Name _____

Telling Time Before and After the Hour

Write the time.

1. 5 minutes after 9 ___:___

2. 20 minutes until 5 ___:___

3. 15 minutes after 6 ___:___

4. 25 minutes before 8 ___:___

5. 30 minutes past 12 ___:___

6. 15 minutes before 11 ___:___

7. 10 minutes after 4 ___:___

8. 40 minutes after 7 ___:___

Write the time in two ways. Use *before* or *after*.

9.

___:___

___ minutes after _____

10.

___:___

___ minutes after _____

11.

___ minutes before _____

12.

___:___

___ minutes _____ _____

13.

___:___

___ minutes _____ _____

14.

___ minutes _____ _____

© Calvert School

Name _____

Fractions and Time

Complete each sentence with a word or number from the box.

1. One quarter of an hour equals _____ minutes.

2. One half hour equals _____ minutes.

3. Three fourths of an hour equals _____ minutes.

4. Another name for 30 minutes after an hour is _____ past.

> half
>
> 45
>
> 15
>
> quarter
>
> 30

Write the times. Use the words *quarter to, quarter past,* or *half past.*

5.

_____

6.

_____

7.

_____

Write each time.

8. quarter to nine ___:___

9. half past six ___:___

10. quarter past two ___:___

11. quarter to eleven ___:___

12. quarter hour before 4:00 ___:___

13. quarter hour after 8:15 ___:___

Solve. Remember to label your answers.

14. Linda worked in her garden for three quarters of an hour. How many minutes did Linda work?

15. When Libby went out to play, her mother told her to come home in half an hour. In how many minutes does Libby need to be home?

MIXED Practice

Add or subtract.

1.   437
      − 98

2.   1,742
      + 6,159

3.   3,617
      − 1,406

4. $6 + (8 + 7) =$ _____

5. _____ $= 28 + 42$

6. $19 − 13 =$ _____

7. Shannon has 8 books, Olivia has 3 puzzles, and Sean has 8 books. How many books are there?

8. If Karen has 3 one-dollar bills, 2 quarters, 4 dimes, 2 nickels, and 18 pennies in her purse, how much money does she have altogether?

Compare. Write <, >, or =.

9. $24 + 33 + 17$ ◯ $28 + 23 + 37$

10. $307 − 29$ ◯ $618 − 459$

Name _____

Elapsed Time

Use a clock to tell how much time has passed from...

1. 10:30 until 11:05   _____

2. 4:05 until 4:40   _____

3. 6:50 until 7:10   _____

4. 8:05 until 9:30   _____

5. 10:02 until 10:32   _____

6. 1:10 until 4:10   _____

Write the time 15 minutes *before* each time shown.

7.    8.    9.    10.

___:___   ___:___   ___:___   ___:___

Solve. Remember to label your answers.

11. Pat turned on her TV at 6:17. At 6:30 her mother called her for dinner. How long did Pat watch TV?

_____

12. Jeremy went out to play at 3:20. His mother told him to be home in 2 hours. What time should Jeremy be home?

_____

The blue hands show the start time. The gray hands show the end time. Circle the letter that shows the elapsed time.

13.
   a. 1 hour
   b. 30 minutes
   c. 3 minutes

14.
   a. 1 hour
   b. 50 minutes
   c. 5 minutes

15. `12:00` to `12:20`

   a. 20 minutes
   b. 40 minutes
   c. 1 hour

16. `3:30` to `4:10`

   a. 30 minutes
   b. 40 minutes
   c. 10 minutes

Circle the letter that shows a reasonable elapsed time for each activity.

17. a birthday party
   a. 20 minutes
   b. 2 hours
   c. 10 hours

18. riding your bike
   a. 30 minutes
   b. 30 hours
   c. 300 hours

MIXED Practice

Solve. Remember to label your answers.

1. The library collected $8.70 in fines on Friday and $9.85 on Saturday. How much did they collect in all?

   _____

2. The cost to park at a parking meter is 10¢ for every 15 minutes. How much would it cost to park for 1 hour?

   _____

3. Write *sixteen thousand, nine hundred five* in standard form.

   _____

4. What comes next in the pattern?

   3rd, 5th, 7th, 9th, _____, _____

Name _____

Reading a Schedule

The chart shows two morning bus schedules from Fairfield to Crownsville.

Bus Schedule

Town	Bus 1 Leave	Bus 2 Leave
Fairfield	7:02	8:58
Greenville	7:18	9:12
Porter	7:38	9:26
Avalon	7:58	9:42
Crystal Lake	8:14	9:57
Crownsville (Arrive)	8:27 Arrive	10:07 Arrive

1. How much earlier does Bus 1 leave Fairfield than Bus 2?

   _____

2. How long does it take Bus 1 to travel from Fairfield to Greenville?

   _____

3. On Bus 2, how long is the trip from Greenville to Porter?

   _____

4. How long does it take Bus 2 to travel from Fairfield to Greenville?

   _____

5. If you had to meet a friend in Crystal Lake at 8:45, which bus should you take?

   _____

6. What time does Bus 1 arrive in Crownsville?

   _____

7. How long does Bus 2 take to get from Fairfield to Crownsville?

   _____

8. How long does Bus 1 take to get from Fairfield to Crownsville?

   _____

Name _____

Problem-Solving Skill: Using a Calendar

September						
S	M	T	W	T	F	S
		1	2	3	4	5
6	7	8	9	10	11	12
13	14	15	16	17	18	19
20	21	22	23	24	25	26
27	28	29	30			

October						
S	M	T	W	T	F	S
				1	2	3
4	5	6	7	8	9	10
11	12	13	14	15	16	17
18	19	20	21	22	23	24
25	26	27	28	29	30	31

November						
S	M	T	W	T	F	S
1	2	3	4	5	6	7
8	9	10	11	12	13	14
15	16	17	18	19	20	21
22	23	24	25	26	27	28
29	30					

Write each date or number of days. Use the calendars.

1. 3 weeks after October 22

   _____

2. the number of days from October 19 to November 9

   _____

3. 10 days after September 19

   _____

4. 10 days before October 12

   _____

5. 5 weeks before November 24

   _____

6. 4 weeks after September 5

   _____

7. the number of days from October 2 to October 23

   _____

8. the number of days from October 17 to November 4

   _____

9. 5 weeks after September 25

   _____

10. the number of days from September 14 to September 29

    _____

11. 4 weeks after November 2

    _____

12. 3 weeks and 4 days before October 26

    _____

September						
S	M	T	W	T	F	S
		1	2	3	4	5
6	7	8	9	10	11	12
13	14	15	16	17	18	19
20	21	22	23	24	25	26
27	28	29	30			

October						
S	M	T	W	T	F	S
				1	2	3
4	5	6	7	8	9	10
11	12	13	14	15	16	17
18	19	20	21	22	23	24
25	26	27	28	29	30	31

November						
S	M	T	W	T	F	S
1	2	3	4	5	6	7
8	9	10	11	12	13	14
15	16	17	18	19	20	21
22	23	24	25	26	27	28
29	30					

Solve. Use the calendars. Remember to label your answers.

13. A bus tour began the morning of November 4 and ended the morning of November 8. How many days did the tour last?

_____

14. The fifth week of school started October 5th. When did the first week start?

_____

15. Mrs. Nicolls is going on a three week trip beginning November 5th. When will she return?

_____

16. Rachel sells girl scout cookies. Her order needs to be turned in 2 weeks after October 14. What date will she turn in her order?

_____

MIXED Practice

Write the elapsed time.

1.

    Start   End

_____

2.

    Start   End

_____

3.

    Start   End

_____

4.

    Start   End

_____

Name _____

Chapter 4 Review

Write the amount of money.

1.

2.

_____ ¢ $ _____ . _____

$ _____ . _____

Write the letter of the group that shows an equal amount.

_____ 3.    a.

_____ 4.    b.

_____ 5.    c.

Write the amount of change.

Cost	Amount Given	Amount of Change
6. 		_____
7. $1.29		_____

Add or subtract.

8.	$1.89	9.	$14.89	10.	$8.07	11.	$56.04
	3.46		+ 27.56		− 3.42		− 19.27
	+ 5.21						

Solve. Remember to label your answers.

12. Ted paid $6.50 for a movie ticket, $4.25 for popcorn, and $2.50 for a drink. How much did he spend in all?

_____

13. Stephanie has 3 quarters, 2 dimes, and 2 nickels. She buys a milkshake for $1.04. How much money does she have left?

_____

Write the time.

14.

_____:_____

15.

_____:_____

16.

Write the time using the words *quarter to, quarter past,* **or** *half past.*

17.

_____

18.

_____

19.

_____

Name _____

Chapter 4 Test Prep

Ring the letter of the correct answer.

1. One $5 bill, one $1 bill, 3 quarters, and 1 nickel equal

   a. $7.80   **b.** $6.80

   c. $6.70   **d.** $5.80

2. $8.68 equals

   a. 8 dollars, 2 quarters, 1 dime, 1 nickel, 3 pennies

   b. 8 dollars, 3 quarters, 1 dime, 2 nickels, 3 pennies

   c. 8 dollars, 1 quarter, 3 dimes, 1 nickel, 3 pennies

   d. 8 dollars, 2 quarters, 2 dimes, 0 nickels, 3 pennies

3. $3.72 + $5.46 equals

   a. $9.28   **b.** $9.18

   c. $8.28   **d.** $8.18

4. $ 10.08 − $8.79 equals

   a. $1.39   **b.** $1.29

   c. $1.27   **d.** $1.19

5. Connie wants to buy a new CD that costs $9.87.
   So far, she has saved $6.48. How much more money does she need?

   a. $3.39   **b.** $3.41

   c. $2.48   **d.** $2.39

6. What time is shown on the clock?

    a. 3:35   **b.** 4:35

    c. 7:18   **d.** 7:20

7. What time will it be in 40 minutes?

    a. 12:00   **b.** 11:55

    c. 11:20   **d.** 11:04

8. How do you write the time in words?

    a. 35 minutes after 6

    b. 29 minutes before 8

    c. 22 minutes before 7

    d. 31 minutes after 8

9. How do you write the time *quarter to 10*?

    a. 9:30   **b.** 9:45

    c. 10:15   **d.** 10:45

10. Adam's train was due in the station at 2:20. It did not arrive until 3:08. How many minutes late was the train?

    a. 58 minutes late

    b. 53 minutes late

    c. 48 minutes late

    d. 43 minutes late

Name _____

Collecting and Recording Data

Use the chart to answer the questions.

1. Joe took a survey to find the different ways students carry their books. The frequency table shows the data he gathered. Complete the chart.

2. How many students did Joe survey?

   _____

3. How do most students in this group carry their books?

   _____

Carrying Books

How Books are Carried	Tally	Frequency
Shoulder pack	IIII	
Backpack	HHL HHL I	
Hand-held pack		2
No pack		4

4. What is the least favorite way students carry their books?

   _____

Ellen surveyed the girls in her scout troop to find their favorite color.

5. Record the results from the survey in the frequency table.

Favorite Colors

pink, green, pink, purple, blue, blue, blue, pink, pink, green, pink, yellow, blue, blue, pink, green, purple, pink, pink, pink

Favorite Colors

Color	Tally	Frequency

6. How many scouts did Ellen survey? _____

7. Ellen wants to buy balloons for her scout troop. She can only buy one color. Which color balloon should she choose? Use the data to explain your answer.

_____

_____

MIXED Practice

Add or subtract.

1. $\begin{array}{r} 8,290 \\ - 5,137 \\ \hline \end{array}$ 2. $\begin{array}{r} 4,836 \\ + 1,234 \\ \hline \end{array}$ 3. $\begin{array}{r} \$9.27 \\ - \ 4.08 \\ \hline \end{array}$ 4. $\begin{array}{r} \$5.68 \\ - \ 3.42 \\ \hline \end{array}$

5. _____ $= 6 + (8 + 9)$ 6. _____ $= 85 - 40$

Solve. Remember to label your answers.

7. Grant earned 6 quarters, 10 dimes, 15 nickels, and 75 pennies from selling lemonade. How much money did he earn in all? _____

8. A movie started at 7:10 and ended at 10:35. How long was the movie?

_____

Name _____

Line Plots

Reminder: A line plot uses Xs on a number line to show how often an event occurs or how many of something there are.

Use the data to complete the Neighborhood Trees line plot.

1. As part of a scout project, Ben counts the number of trees in each yard of his neighborhood. Ben found the following numbers of trees:

   1 0 5 4 3 1 0 6 4 2 7 3 4 4 5 6 0 3

   Put the data in order from least to greatest and complete Ben's line plot.

   ## Neighborhood Trees

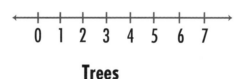

   Trees

Use the data to complete the Ages of Ballet Class Students line plot.

2. Mrs. Anderson wants to set up her spring ballet classes according to the ages of the students. The students enrolled were the following ages:

    7 5 12 15 8 6 5 9 11 10 5 6 6 8 10 13 12

    Put the data in order from least to greatest and complete Mrs. Anderson's line plot.

Ages of Ballet Class Students

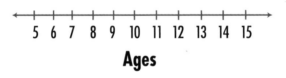

Ages

3. How many students have enrolled? _____

4. What is the age of the youngest student enrolled? _____

5. What is the age of the oldest student enrolled? _____

6. If one class will be for students who are 5 and 6 years old, how many students will be in the class? _____

7. If one class will be for students 12 years of age and older, how many students will be in the class?_____

Name _____

Mode and Range

The number that occurs the most often in a set of data is called the **mode**.

The **range** is the difference between the greatest number and the least number in a set of data.

The **median** is the middle number when the numbers are written in order.

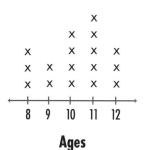

Ages of Boys in the Chorus

Ages

Use the line plot on Ages of Boys in the Chorus to answer the following questions.

1. What is the mode of the data? _____

2. What is the range of the data? _____

3. Explain how you found the range.

   _____

   _____

   _____

4. What is the median of the data? _____

5. Two eleven-year-old boys left the chorus. What is the new mode?

   _____

6. Does the range change? Explain.

   _____

   _____

Use the line plot on the Height of Girls on the Basketball Team to answer the following questions.

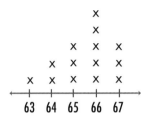

Height of Girls on the Basketball Team

Inches

7. What is the mode of the data?

_____

8. What is the range of the data?

_____

9. If a girl who is 68 inches tall joins the team, what is the new range?

_____

10. Will the new player's height affect the mode? Why or why not?

_____

_____

_____

Name _____

Pictographs

Glen gathered data about the number of animals at the local zoo. He recorded the data in the chart.

1. **a.** Use Glen's data to make a pictograph. Decide on a symbol that is easy to draw. Let each symbol represent 1 animal. Make a key. Complete the pictograph.

   b. How many more lions than elephants are there?

   _____

   c. How many animals are there altogether?

   _____

   d. Are there more bears and lions combined or elephants and giraffes combined?

   _____

Animals in the Zoo

Animal	Number
Bears	5
Elephants	3
Giraffes	6
Lions	8

Animals in the Zoo

Animal	Number

Key: _____ = 1 animal

Andrea counted the kinds of lunches that were purchased at the refreshment stand. She recorded the data in a chart.

Lunches Bought by Visitors

Lunch	Number
Hot dog	15
Pizza slice	35
Hamburger	20
Chicken nuggets	10

2. **a.** Use Andrea's data to make a pictograph. Decide on a symbol that is easy to draw. Let each symbol represent 5 lunches. Make a key and complete the pictograph.

b. Which kind of lunch was the most popular?

_____

c. How many fewer chicken nugget lunches were bought than pizza slices?

Lunches Bought by Visitors

Lunch	Number

Key: _____ = 5 lunches

_____

d. How many more pizza slices were bought than hot dogs?

_____

e. Andrea drew 15 symbols next to *Hot dog* on her pictograph. Explain Andrea's mistake.

_____

_____

_____

Name _____

Bar Graphs

In a *vertical* bar graph, the bars go up and down. The same information can be shown in a *horizontal* bar graph, where the bars go from left to right.

Weekday Noon Temperatures

Day	Temp. at Noon
Monday	26°C
Tuesday	24°C
Wednesday	20°C
Thursday	17°C
Friday	15°C

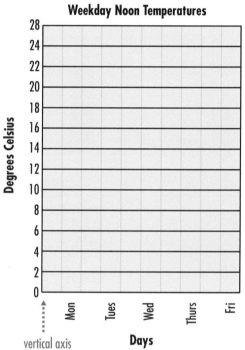

1. Color the bars to show the data in the chart. Since the numbers are on the *vertical axis,* the bars will go up.

2. Now use the same data to create a *horizontal* bar graph. Notice that the numbers are on the *horizontal axis,* so the bars will go from left to right.

The graphs show the same data in two ways.

3. Compare the graphs on this page. How are they alike and how are they different?

_____

_____

_____

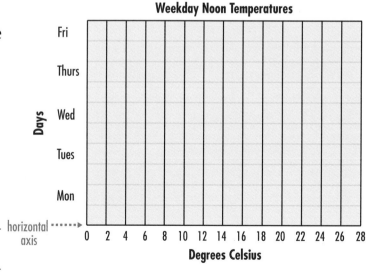

The bar graph shows students' favorite ice cream flavors.

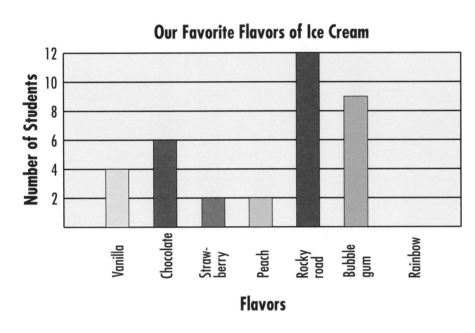

Use the vertical bar graph to answer each question.

1. How many students chose each flavor as their favorite?

   a. peach _____   **b.** rocky road _____   **c.** bubble gum _____

   d. vanilla _____   **e.** chocolate _____   **f.** strawberry _____

   g. rainbow _____

2. How many more students chose chocolate than peach as their favorite? _____

3. How many votes did chocolate and vanilla receive altogether? _____

4. Which flavor was liked by more of the students than any other flavor? _____

Name _____

5. Use the data in the table to make a horizontal bar graph.
Follow the steps below.

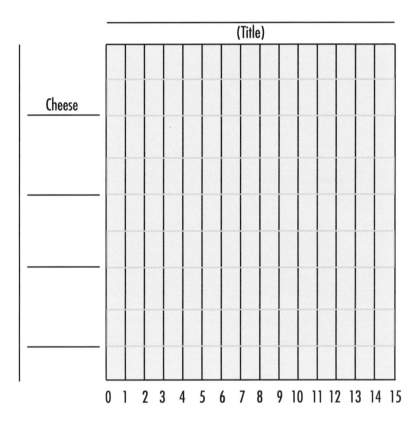

Favorite Pizza	
Cheese	10
Pepperoni	14
Sausage	5
Anchovy	1

a. Write the title on the line above the graph.

b. Label the line on the side of the graph "Kinds of Pizza."

c. Label the line below the graph "Number of Votes."

d. Write the types of pizza beside each bar.

e. Color each bar to display the data in the table.

6. Without looking at the numbers, how can you tell which type of pizza is liked by the fewest number of people? Explain your answer.

_____

_____

Name _____

Problem-Solving Strategy: Making a Tree Diagram

Make a tree diagram to solve each problem.

1. For his birthday, Lance received two pairs of pants, one pair of blue and one pair of black. He also received three shirts—white, yellow, and red. List the different shirt and pants combinations Lance can make.

   Tree   Combinations

2. Ice cream cones are being sold at a festival. You can get a cake cone or a sugar cone. You can get a scoop of vanilla or chocolate ice cream. List the different combinations of one-scoop cones you can buy.

   Tree   Combinations

3. Healthy Sub serves four types of sandwiches— turkey, ham, tuna, and veggie. There are three types of bread— wheat, white, and rye. The possible toppings are mustard and mayo. List all the different bread, filling, and 1-topping sandwich combinations.

    Tree   Combinations

Name _____

Graphing Ordered Pairs

To show locations on a grid,
count *over* first, then *up* ↑.

Write the coordinates for each point.

1. C _____   2. A _____

3. G _____   4. E _____

Name the point with the given coordinates.

5. (2, 8) _____   6. (3, 3) _____

7. (5, 9) _____   8. (8, 3) _____

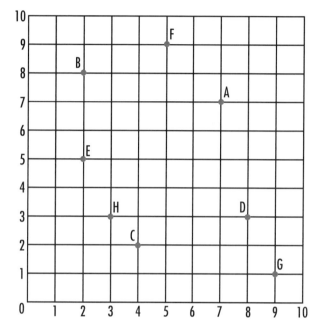

Plot and label each point on the grid.

9. L (5, 6)   10. M (0, 9)

11. N (9, 2)   12. O (6, 2)

13. P (10, 0)   14. Q (2, 1)

15. R (8, 7)   16. S (5, 10)

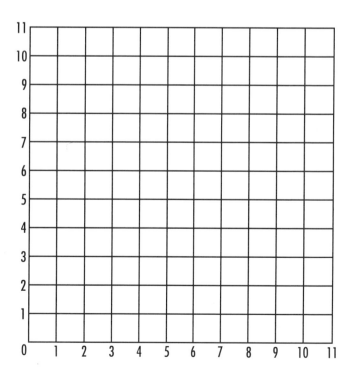

© Calvert School

Name _____

Probability and Predicting Outcomes

Materials

    15 cubes, some of each of 4 different colors
    paper bag
    a helper
    paper and pencil

Ask your helper to put 15 cubes in the bag.
There should be 4 different colors.

1. Without looking, draw a
   cube from the bag. Record
   the color in the frequency
   table.

Cube Colors

Color	Tally	Frequency

_____

2. Place the cube back in the bag. Draw again and record the color.

_____

3. Repeat steps 1 and 2 until you have drawn a cube from the bag 25
   times.

_____

4. Using the data you gathered, predict how many cubes of each color
   are in the bag. Record your predictions.

_____

5. Look at the cubes in the bag. Count each color. How close were your
   predictions to the actual amounts?

_____

Use the data to make predictions.

Every week Marcie takes a quiz to find out how many addition facts she can answer correctly in 4 minutes. There are 100 facts on each quiz. Her results are shown in the chart.

Addition Fact Scores

Week	1	2	3	4	5
Number Correct	76	74	81	92	

6. Predict Marcie's score for Week 5.

    a. 105   **b.** 96   **c.** 75

Explain. _____

MIXED Practice

Write the time. Use A.M. or P.M.

1. 17 minutes after 4:15 A.M.

    _____

2. 33 minutes after 6:12 P.M.

    _____

3. Jen was eating breakfast at 10 minutes after 7 in the morning. Write the time.

    _____

Add or subtract.

4. $\begin{array}{r} 367 \\ + 721 \\ \hline \end{array}$ 5. $\begin{array}{r} 419 \\ - 125 \\ \hline \end{array}$ 6. $\begin{array}{r} 48 \\ + 29 \\ \hline \end{array}$

7. _____ = 23 + 46   8. _____ = 91 + 5

Name _____

Circle Graphs

A *circle graph* is sometimes used to show parts of a whole.
A circle graph looks like a pie that has been cut into slices. A large slice represents a large amount and a small slice represents a small amount.

Favorite Movies

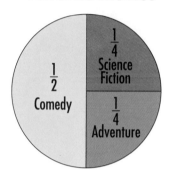

Haley asked the children at her family reunion what type of movies they prefer. She displayed the results in a circle graph.

1. What kind of movie do most of the children prefer?

_____

2. How do you know?

_____

The circle graph shows how Jake spends his $10 allowance.

Jake's Allowance

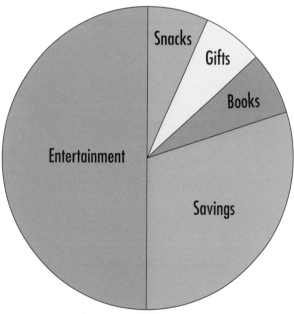

3. On what does Jake spend most of his money?

_____

4. What does he do with the second greatest amount of money?

_____

5. For what things does he spend an equal amount?

_____

6. What does the whole circle represent?

_____

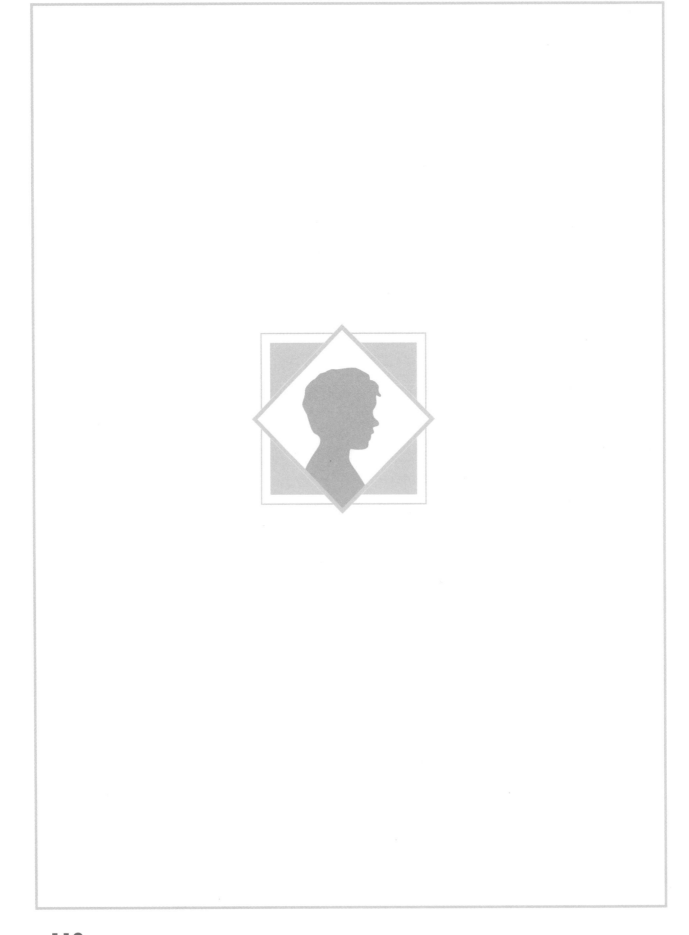

Name _____

Chapter 5 Review

1. The data in the box shows answers to the question, "What is your favorite subject?" Use the data to complete the frequency table.

reading, spelling, math, math, science, spelling, science, reading, reading, math, math, science, spelling, science, reading, math, reading, math, science

Favorite Subjects

Favorite Subject	Tally	Number of Votes
Reading		
Spelling		
Science		
Math		

2. Make a vertical bar graph to display the data in the frequency table from problem 1. Remember to label the parts of the graph.

(Title)

6
5
4
3
2
1
0

Answer the questions about Anika's reading scores.

Anika's Reading Scores

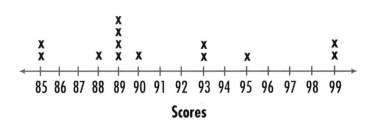

Scores

3. What are the median, mode, and range of the data?

median _____ mode _____ range _____

4. Make a tree diagram to show the possible combinations for an ice cream sundae at Busby's Ice Cream Shoppe. You can choose one ice cream flavor from chocolate, vanilla, or strawberry and one topping of hot fudge or butterscotch.

Tree   Combinations

Plot and label each point on the grid.

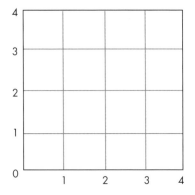

5. A (1, 1)   6. B (3, 2)

7. C (2, 4)   8. D (1, 3)

Name _____

Chapter 5 Test Prep

Ring the correct answer.
Use the bar graph to answer the questions.

Classroom Pets

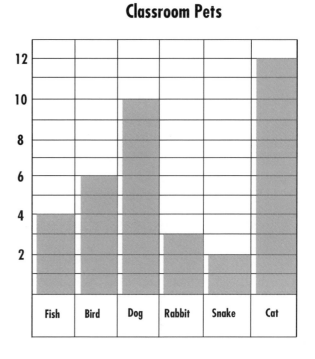

1. How many students have a pet rabbit?

   a. 2   b. 3
   c. 4   d. 6

2. How many more students have cats than fish?

   a. 4   b. 6
   c. 7   d. 8

3. Which is the least popular pet?
   a. rabbit   b. fish
   c. snake   d. cat

4. How many students have birds, dogs, and snakes altogether?
   a. 18   b. 17
   c. 16   d. 15

Use the line plot to answer the questions.

The line plot shows the weekly allowances received by students in Grade 3.

5. What is the allowance most frequently received?
    a. $1.00   b. $2.00
    c. $1.50   d. $2.50

6. What is the range of the data?
    a. $1.00   b. $1.50
    c. $2.00   d. $3.00

7. What is the mode of the data?
    a. $1.00   b. $2.00
    c. $1.50   d. $3.00

Students' Weekly Allowances

Allowances

Ring the letter of the correct answer.

8. A pictograph is like
    a. a bar graph with pictures.   b. a coordinate grid with pictures.
    c. a tree diagram with pictures.   d. a circle graph.

9. To show points on a grid, you should
    a. draw vertical bars.
    b. divide a circle into different parts.
    c. plot the coordinates and label them.
    d. make a pictograph.

10. A tree diagram is used to show
    a. the parts of a tree.
    b. all the possible answer combinations to a problem.
    c. some of the possible answer combinations to a problem.
    d. a drawing of the problem.

Name _____

Exploring the Meaning of Multiplication

Write a multiplication equation for each picture.

1.   ____ × ____ = ____

2.

    ____ × ____ = ____

3.

    ____ × ____ = ____

4.

    ____ × ____ = ____

5.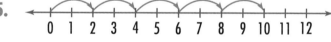

    _____

6.

    _____

7.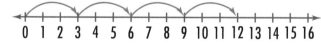

    _____

8.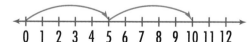

    _____

Write a multiplication equation for each.

9. $3 + 3$ **10.** $4 + 4 + 4 + 4$ **11.** $5 + 5 + 5$

_____   _____   _____

Draw a picture to show each multiplication equation.

12. $3 \times 6 = 18$ **13.** $5 \times 4 = 20$

MIXED Practice

Add or subtract.

1.	315 + 425	2.	561 − 180	3.	9,803 − 2,694

4.	7,401 + 989	5.	$48.54 − 19.25	6.	$15.38 + 17.21

Solve. Remember to label your answer.

7. Tony is eighth in line. Celia is three places in front of Tony. What place is Celia in line?

_____

Name _____

Multiplication Properties: Commutative, Zero, and Identity

Use the words *factors* or *product* to label the numbers.

1. $9 \times 2 = 18$
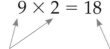

_____   _____

2.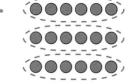

$$\begin{array}{r} 9 \\ \times\, 2 \\ \hline 18 \end{array}$$

Write a multiplication equation for each picture.

3.

____ × ____ = ____

____ × ____ = ____

4.

____ × ____ = ____   ____ × ____ = ____

Complete the statements.

5. When one factor is zero, the product is _____.

6. The product of one times a number is _____.

Write the missing numbers.

7. $8 \times$ _____ $= 0$

8. $9 = 9 \times$ _____

9. $1 \times$ _____ $= 6$

10. $0 =$ _____ $\times 6$

11. _____ $\times 5 = 5$

12. _____ $= 1 \times 0$

Grade 3 • Chapter 6   **119**

© Calvert School

Multiply.

13. $3 \times 6 =$ _____

$6 \times 3 =$ _____

14. _____ $= 3 \times 7$

_____ $= 7 \times 3$

15. $1 \times 8 =$ _____

$8 \times 1 =$ _____

16.
$\begin{array}{cc} 5 & 3 \\ \times 3 & \times 5 \\ \hline \end{array}$

17.
$\begin{array}{cc} 0 & 3 \\ \times 3 & \times 0 \\ \hline \end{array}$

18.
$\begin{array}{cc} 6 & 2 \\ \times 2 & \times 6 \\ \hline \end{array}$

Write *true* or *false*.

19. $4 \times 2 = 2 \times 3$

_____

20. $5 \times 4 = 4 \times 5$

_____

21. $9 \times 2 = 2 \times 9$

_____

22. $3 \times 1 = 1 \times 3$

_____

23. $8 \times 0 = 1 \times 8$

_____

24. $0 \times 9 = 0 \times 5$

_____

Solve. Write a multiplication equation for each problem. Remember to label your answers.

25. One dog has 1 tail. How many tails do 4 dogs have?

_____

26. Each basket has 0 apples. How many apples are in 1 basket?

_____

MIXED Practice

Write the place value of the 6 in each number.

1. 5,561 _____   **2.** 4,613 _____   **3.** 26,117 _____

Name _____

Arrays

You can draw an **array** to show multiplication.
An array has equal rows.
This array shows 5×4.

THINK
5 rows with 4 dots in each row

Draw an array to find each product. Be sure to line up your dots so they are in straight rows and columns.

1. $2 \times 3 =$ __ **2.** $5 \times 3 =$ __ **3.** $4 \times 1 =$ __ **4.** $2 \times 6 =$ __

  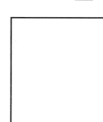 

Write a multiplication equation for each array.

5.   **6.**   **7.**

___ $\times$ ___ = ___   ___ $\times$ ___ = ___   ___ $\times$ ___ = ___

Write the factors and product for each array.

8. **9.** **10.**

Factors ____ , ____   Factors ____ , ____   Factors ____ , ____

Product _____   Product _____   Product _____

Name _____

Multiplying by 2

Multiply.

1. $0 \times 2 =$ _____ $1 \times 2 =$ _____

   $2 \times 2 =$ _____ $3 \times 2 =$ _____

   $4 \times 2 =$ _____ $5 \times 2 =$ _____

   $6 \times 2 =$ _____ $7 \times 2 =$ _____

   $8 \times 2 =$ _____ $9 \times 2 =$ _____

2. Find each product. Below each problem, write the letter that matches the product. You will spell two mystery words.

$$
\begin{array}{ccccccccccc}
8 & 5 & 2 & 4 & 2 & 9 & 7 & 6 & 8 & 3 & 2 \\
\times 2 & \times 2 & \times 8 & \times 2 & \times 4 & \times 2 & \times 2 & \times 2 & \times 2 & \times 2 & \times 2 \\
\hline
\end{array}
$$

___ ___ ___ ___   ___ ___ ___ ___ ___ ___ ___

16 = E   18 = U   6 = R   10 = V   4 = S   8 = N   12 = B   14 = M

3. How are the words you spelled in problem 2 related to the products you found in problem 1?

   _____

Multiply.

4. $8 \times 2 =$ _____   5. _____ $= 1 \times 2$ 6. $2 \times 7 =$ _____

7. _____ $= 3 \times 2$ 8. $2 \times 9 =$ _____   9. _____ $= 2 \times 5$

Name _____

Multiplying by 5

Multiply.

1. $0 \times 5 = $ _____ $1 \times 5 = $ _____

   $2 \times 5 = $ _____ $3 \times 5 = $ _____

   $4 \times 5 = $ _____ $5 \times 5 = $ _____

   $6 \times 5 = $ _____ $7 \times 5 = $ _____

   $8 \times 5 = $ _____ $9 \times 5 = $ _____

2. Multiply the number in the center by each number in the inside ring. Write the products in the outside ring.

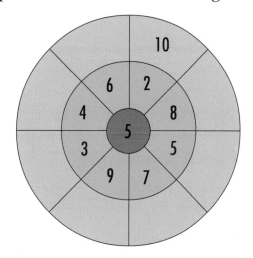

Multiply.

3. $5 \times 2 = $ _____   4. _____ $= 5 \times 9$ 5. $3 \times 5 = $ _____

6. $\begin{array}{r} 6 \\ \times 5 \\ \hline \end{array}$ 7. $\begin{array}{r} 7 \\ \times 5 \\ \hline \end{array}$ 8. $\begin{array}{r} 5 \\ \times 1 \\ \hline \end{array}$ 9. $\begin{array}{r} 5 \\ \times 0 \\ \hline \end{array}$ 10. $\begin{array}{r} 8 \\ \times 5 \\ \hline \end{array}$

MIXED Practice

Write the amount of money.

11. 4 nickels = _____¢ 12. _____¢ = 9 nickels 13. 6 nickels = _____¢

14. 3 nickels = _____¢ 15. _____¢ = 5 nickels 16. 8 nickels = _____¢

Solve. Remember to label your answers.

1. Describe the pattern 13, 17, 21, 25.

    _____

2. Tanya has 2 one-dollar bills, 7 quarters, 5 dimes,
    10 nickels, and 6 pennies. How much money
    does she have?

    _____

3. A bookcase has 26 books on the first shelf, 18 on the
    second shelf, and 24 on the third shelf. How many books
    are there in all?

    _____

4. Write the time in two ways.

    _____:_____   _____ after _____

Name _____

Hidden Name

Solve each problem. Draw an array or use repeated addition to help you. Then shade a box for each product.

You will reveal the name of the Super Hero of Repeated Addition.

0	1	2	3	4	5	6	7	8	9	10	11	12	13	14	15	16	17	18	19	20
21	22	23	24	25	26	27	28	29	30	31	32	33	34	35	36	37	38	39	40	41
42	43	44	45	46	47	48	49	50	51	52	53	54	55	56	57	58	59	60	61	62
63	64	65	66	67	68	69	70	71	72	73	74	75	76	77	78	79	80	81	82	83
84	85	86	87	88	89	90	91	92	93	94	95	96	97	98	99	100	101	102	103	104

1. $2 \times 2 =$	**2.** $19 \times 0 =$	**3.** $26 \times 2 =$
4. $15 \times 5 =$	**5.** $96 \times 1 =$	**6.** $33 \times 2 =$
7. $32 \times 2 =$	**8.** $51 \times 1 =$	**9.** $7 \times 2 =$
10. $11 \times 2 =$	**11.** $7 \times 5 =$	**12.** $1 \times 49 =$
13. $56 \times 1 =$	**14.** $12 \times 5 =$ (Hint: Look at a clock!)	**15.** $14 \times 2 =$
16. $42 \times 2 =$	**17.** $2 \times 4 =$	**18.** $1 \times 73 =$
19. $2 \times 6 =$	**20.** $81 \times 1 =$	**21.** $5 \times 4 =$
22. $93 \times 1 =$	**23.** $8 \times 5 =$	**24.** $1 \times 77 =$
25. $1 \times 33 =$	**26.** $44 \times 2 =$	**27.** $9 \times 1 =$
28. $1 \times 54 =$	**29.** $35 \times 2 =$	**30.** $1 \times 92 =$
31. $8 \times 2 =$	**32.** $38 \times 1 =$	**33.** $2 \times 12 =$
34. $49 \times 2 =$	**35.** $97 \times 1 =$	**36.** $2 \times 51 =$
37. $2 \times 22 =$		

38. What is the name of the Super Hero of Repeated Addition?

_____

Name _____

Problem-Solving Strategy: Acting It Out or Making a Model

Solve. Make a model or use counters to act out each problem. Remember to label your answers.

1. Ella bought 12 goldfish. She divided them equally among 4 fish bowls. How many goldfish are in each bowl?

2. David organized his CD collection into baskets. He put 5 CDs into each basket. If he filled 8 baskets, how many CDs does he have?

3. Darren, Joe, Stan, and Will went fishing. Each boy caught 4 fish. How many fish did the boys catch altogether?

4. Devon's parents bought 3 pizzas. If there were 8 slices in each pizza, how many slices were there in all?

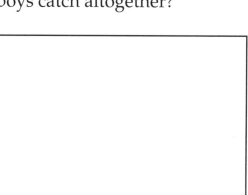
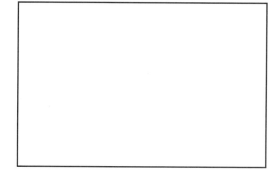

5. Maddie made muffins for a bake sale. She placed 6 muffins in 1 tin. If Maddie filled 6 tins, how many muffins did she make in all?

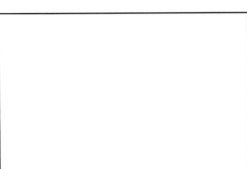

6. Dennis is going shopping. He has 7 five-dollar bills. How much money does Dennis have in all?

7. Mr. Mattas asked his students to line up in equal rows. There are 24 students in his class. Show the different ways the students could line up.

8. List the factors of 24.

_____

Name _____

Multiplying by 10

Multiply.

1. $2 \times 10 =$ ___

2. $10 \times 1 =$ ___

3. $0 \times 10 =$ ___

4. $4 \times 10 =$ ___

5. ___ $= 10 \times 10$

6. $10 \times 3 =$ ___

7. $8 \times 10 =$ ___

8. $10 \times 9 =$ ___

9. ___ $= 5 \times 10$

10. ___ $= 10 \times 6$

11. ___ $= 10 \times 7$

12. $10 \times 10 =$ ___

Write *true* or *false*.

13. $10 \times 8 > 79$ _____

14. $78 > 10 \times 7$ _____

15. $10 \times 10 = 10$ _____

16. $101 < 10 \times 10$ _____

17. $10 \times 5 < 65$ _____

18. $30 = 10 \times 4$ _____

19. Multiply the number in the center by each number in the inside ring.
    Write the products in the outside ring.

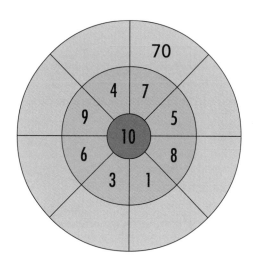 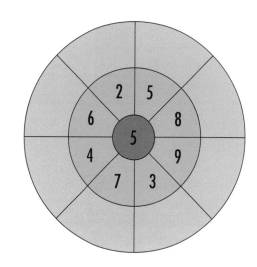

Name _____

Multiply by 10 Puzzle

Look at each multiplication equation.

$5 \times 10 = 50$ $\qquad$ $6 \times 10 = 60$ $\qquad$ $7 \times 10 = 70$

You can see a pattern. When you multiply by 1 ten your answer ends in zero. 5 times 10 is 5 tens or 50.

The same thing happens when you multiply greater numbers by 10.

Look for the pattern:

$12 \times 10 = 12\underline{0}$ $\qquad$ $120 \times 10 = 1,20\underline{0}$ $\qquad$ $1,200 \times 10 = 12,00\underline{0}$

Extend your skills with multiplying by 10. Solve the problems and write the answers in the crossword puzzle. Keep in mind the pattern above.

Across
1. $291 \times 10 =$ _____
3. $129 \times 10 =$ _____
5. $10 \times 22 =$ _____
7. $451 \times 10 =$ _____
8. $10 \times 75 =$ _____
9. $810 \times 10 =$ _____

Down
2. $10 \times 97 =$ _____
3. $17 \times 10 =$ _____
4. $9,342 \times 10 =$ _____
5. $20 \times 10 =$ _____
6. $548 \times 10 =$ _____
8. $72 \times 10 =$ _____

Name _____

Multiplying by 3

Multiply.

1. $0 \times 3 =$ _____ $1 \times 3 =$ _____

    $2 \times 3 =$ _____ $3 \times 3 =$ _____

    $4 \times 3 =$ _____ $5 \times 3 =$ _____

    $6 \times 3 =$ _____ $7 \times 3 =$ _____

    $8 \times 3 =$ _____ $9 \times 3 =$ _____

2. Count by threes. Write the missing numbers.

Show the multiplication problem on the number line.

3. 7×3

$\overset{\longleftrightarrow}{}$
0 1 2 3 4 5 6 7 8 9 10 11 12 13 14 15 16 17 18 19 20 21 22

4. 3×6

$\overset{\longleftrightarrow}{}$
0 1 2 3 4 5 6 7 8 9 10 11 12 13 14 15 16 17 18 19 20 21 22

Draw a picture to show each problem. Write the product.

5. $5 \times 3 =$ _____

6. $3 \times 9 =$ _____

Multiply.

7. $8 \times 3 =$ _____   8. _____ $= 3 \times 7$ 9. $4 \times 3 =$ _____

10. $\begin{array}{r} 3 \\ \times 6 \\ \hline \end{array}$ 11. $\begin{array}{r} 3 \\ \times 2 \\ \hline \end{array}$ 12. $\begin{array}{r} 3 \\ \times 0 \\ \hline \end{array}$ 13. $\begin{array}{r} 1 \\ \times 3 \\ \hline \end{array}$ 14. $\begin{array}{r} 9 \\ \times 3 \\ \hline \end{array}$

15. $\begin{array}{r} 8 \\ \times 3 \\ \hline \end{array}$ 16. $\begin{array}{r} 6 \\ \times 3 \\ \hline \end{array}$ 17. $\begin{array}{r} 3 \\ \times 9 \\ \hline \end{array}$ 18. $\begin{array}{r} 4 \\ \times 3 \\ \hline \end{array}$ 19. $\begin{array}{r} 3 \\ \times 5 \\ \hline \end{array}$

MIXED Practice

Solve. Remember to label your answers.

1. Sarah wants to make 36 cupcakes. She needs 2 cups of flour and 1 egg to make 9 cupcakes. How many cups of flour and how many eggs will she need to make 36 cupcakes? Fill in the chart.

Ingredients for Cupcakes				
Cups of flour	2			
Eggs	1			
Cupcakes	9	18	27	36

2. At the Bagel Shop, Michael received $2.85 change from a $5 bill. Use the chart to determine what he bought.

Bagel Shop Prices	
Plain bagel	60¢
Cream cheese	25¢
Soda	45¢

_____

Name _____

Multiplying by 4

Multiply.

1. $0 \times 4 =$ _____ $1 \times 4 =$ _____

   $2 \times 4 =$ _____ $3 \times 4 =$ _____

   $4 \times 4 =$ _____ $5 \times 4 =$ _____

   $6 \times 4 =$ _____ $7 \times 4 =$ _____

   $8 \times 4 =$ _____ $9 \times 4 =$ _____

Show the multiplication problem on the number line.

2. 5×4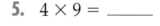
0 1 2 3 4 5 6 7 8 9 10 11 12 13 14 15 16 17 18 19 20 21 22 23

3. 4×3
0 1 2 3 4 5 6 7 8 9 10 11 12 13 14 15 16 17 18 19 20 21 22 23

Draw a picture to show each problem. Write the product.

4. $6 \times 4 =$ _____

5. $4 \times 9 =$ _____

Multiply.

6. $7 \times 4 =$ _____

7. _____ $= 4 \times 8$

8. $2 \times 4 =$ _____

$4 \times 7 =$ _____

_____ $= 8 \times 4$

$4 \times 2 =$ _____

9. $\begin{array}{r} 9 \\ \times\ 4 \\ \hline \end{array}$

10. $\begin{array}{r} 8 \\ \times\ 4 \\ \hline \end{array}$

11. $\begin{array}{r} 4 \\ \times\ 6 \\ \hline \end{array}$

12. $\begin{array}{r} 0 \\ \times\ 4 \\ \hline \end{array}$

13. $\begin{array}{r} 1 \\ \times\ 4 \\ \hline \end{array}$

Solve. Remember to label your answer.

14. How many leaves are on 8 four-leaf clovers? _____

MIXED Practice

Solve. Remember to label your answers.

1. Harry got on the bus at 8:35. He got off of the bus at 9:07. How long was the bus ride?

   _____

2. Tim has 5 coins totaling 61¢. What coins does he have?

   _____

Name _____

Poetry Puzzle

**Multiply to unlock the mystery of this mathematical poem.
Then find the word in the box that matches the answer. Write
each word on the line above the problem.**

0 how	1 me	2 see	3 love	4 ten	5 seven	6 the
8 where	10 multiply	12 five	14 four	15 and	16 to	18 let
20 so	21 times	24 oh	25 just	27 or	30 will	32 I
35 two	36 again	40 run	45 numbers	50 go	60 heart	70 you

_____ ,   _____   _____ ,
$2 \times 5 = \underline{\ \ }$ $3 \times 8 = \underline{\ \ }$ $2 \times 5 = \underline{\ \ }$

_____   _____   _____   _____   _____ ,
$15 \times 0 = \underline{\ \ }$ $4 \times 8 = \underline{\ \ }$ $3 \times 1 = \underline{\ \ }$ $7 \times 10 = \underline{\ \ }$ $4 \times 5 = \underline{\ \ }$

_____   _____   _____   _____
$2 \times 16 = \underline{\ \ }$ $1 \times 3 = \underline{\ \ }$ $4 \times 4 = \underline{\ \ }$ $2 \times 1 = \underline{\ \ }$

_____   _____   _____   _____   _____ .
$2 \times 4 = \underline{\ \ }$ $2 \times 3 = \underline{\ \ }$ $9 \times 5 = \underline{\ \ }$ $6 \times 5 = \underline{\ \ }$ $5 \times 10 = \underline{\ \ }$

_____   _____   _____
$2 \times 7 = \underline{\ \ }$ $7 \times 3 = \underline{\ \ }$ $5 \times 7 = \underline{\ \ }$

_____   _____   _____   _____ ,
$3 \times 9 = \underline{\ \ }$ $2 \times 6 = \underline{\ \ }$ $3 \times 7 = \underline{\ \ }$ $2 \times 2 = \underline{\ \ }$

_____   _____   _____   _____   _____
$6 \times 4 = \underline{\ \ }$ $3 \times 6 = \underline{\ \ }$ $1 \times 1 = \underline{\ \ }$ $5 \times 5 = \underline{\ \ }$ $2 \times 5 = \underline{\ \ }$

_____   _____   _____ .
$4 \times 9 = \underline{\ \ }$ $3 \times 5 = \underline{\ \ }$ $36 \times 1 = \underline{\ \ }$

Name _____

Problem-Solving Application: Using Multiplication

Solve. Remember to label your answers.

1. A butterfly has 6 legs. How many legs do 10 butterflies have?

2. Sue used 4 hexagons in a design. Each hexagon has six corners. How many corners were in her design?

3. At a restaurant, each table has 8 chairs. How many people can sit at 10 tables?

4. There are 12 eggs in 1 carton. How many eggs are there in 2 cartons?

5. You need 2 sticks of butter for one batch of cookies. How many sticks of butter do you need for 3 batches of cookies?

6. What is three groups of one plus four groups of eight?

Name _____

Chapter 6 Review

Write the letter of the correct answer. You may use an answer more than once.

_____ 1. In $9 \times 4 = 36$, 4 is a _____.

_____ 2. If 0 is one of the factors, the product is _____.

_____ 3. If $3 \times 8 = 24$, 24 is the _____.

_____ 4. The answer to a multiplication problem is called the _____.

a. sum

b. factor

c. addend

d. product

e. zero

f. the other factor

Draw a picture to show each problem. Write the product.

5. $5 \times 7 =$ _____

6. $8 \times 3 =$ _____

Multiply.

7. $3 \times 7 =$ _____

$7 \times 3 =$ _____

8. _____ $= 5 \times 8$

_____ $= 8 \times 5$

9. $3 \times 9 =$ _____

$9 \times 3 =$ _____

10. $\begin{array}{r} 4 \\ \times\ 0 \\ \hline \end{array}$

11. $\begin{array}{r} 7 \\ \times\ 2 \\ \hline \end{array}$

12. $\begin{array}{r} 6 \\ \times\ 5 \\ \hline \end{array}$

13. $\begin{array}{r} 8 \\ \times\ 4 \\ \hline \end{array}$

14. $\begin{array}{r} 6 \\ \times\ 3 \\ \hline \end{array}$

15. $\begin{array}{r} 5 \\ \times\,1 \\ \hline \end{array}$ 16. $\begin{array}{r} 8 \\ \times\,2 \\ \hline \end{array}$ 17. $\begin{array}{r} 5 \\ \times\,3 \\ \hline \end{array}$ 18. $\begin{array}{r} 4 \\ \times\,8 \\ \hline \end{array}$ 19. $\begin{array}{r} 7 \\ \times\,1 \\ \hline \end{array}$

Write the missing numbers.

20. $7 \times \underline{\hspace{1cm}} = 7$ 21. $\underline{\hspace{1cm}} \times 9 = 0$ 22. $\underline{\hspace{1cm}} \times 6 = 6$

23. $\underline{\hspace{1cm}} \times 355 = 0$ 24. $925 \times \underline{\hspace{1cm}} = 925$ 25. $5 \times \underline{\hspace{1cm}} = 5$

Solve. Remember to label your answers.

26. A grasshopper has 6 legs. How many legs are on 3 grasshoppers?

27. A bear has 5 claws on each paw. How many claws are on 4 bear paws?

28. A toad has 4 legs. How many legs are on 6 toads?

29. Each page has 3 puzzles. How many puzzles are on 5 pages?

30. A tricycle has 3 wheels. How many wheels are there on 8 tricycles?

31. Tickets to the zoo are $8.00 each. How much will 4 tickets cost?

32. A teacher arranges 12 chairs in equal rows. Draw the different ways he can arrange the chairs.

33. Carla has a sheet of stickers. The sheet has 6 rows of 5 stickers. How many stickers does Carla have?

Name _____

Chapter 6 Test Prep

Ring the letter of the correct answer.

1. Which shows an array?

   a. 🍀🍀🍀 🍀🍀🍀
      🍀🍀🍀

   b. ♥♥♥♥
      ♥♥♥♥
      ♥♥♥♥

   c. $4 + 4 + 4 + 4$

   d. none of these

2. Which lists all the factors of 12?

   a. 1, 2, 3, 4, 6, 12

   b. 12, 24, 36, 48

   c. 2, 4, 6, 8, 12

   d. 3, 6, 9, 12

3. Multiply. $10 \times 4 =$ ____

   a. 4

   b. 20

   c. 44

   d. none of these

4. What is the missing number? $345 \times$ ____ $= 345$

   a. 0

   b. 1

   c. 2

   d. 345

5. Which property is shown? $34 \times 2 = 2 \times 34$

   a. Zero Property of Multiplication

   b. Identity Property of Multiplication

   c. Commutative Property of Multiplication

   d. none of these

6. Multiply. $3 \times 9 =$ ____

   a. 3   b. 21   c. 27   d. 39

7. Multiply. $8 \times 4 =$ ____

   a. 40   b. 36   c. 32   d. 30

8. Multiply. $0 \times 5 =$ ____

   a. 0   b. 1   c. 5   d. none
                                              of these

9. How many sides are on 7 squares? (Hint: A square has 4 sides.)

   a. 21 sides   b. 24 sides   c. 27 sides   d. 28 sides

10. Cookies are 3 for $1.00. How much would 9 cookies cost?

    a. $1.00   b. $3.00   c. $9.00   d. $27.00

Name _____

Multiplying by 6

Multiply.

1. $0 \times 6 =$ _____ $1 \times 6 =$ _____

    $2 \times 6 =$ _____ $3 \times 6 =$ _____

    $4 \times 6 =$ _____ $5 \times 6 =$ _____

    $6 \times 6 =$ _____ $7 \times 6 =$ _____

    $8 \times 6 =$ _____ $9 \times 6 =$ _____

Show each multiplication problem on the number line.

2. $3 \times 6 =$ _____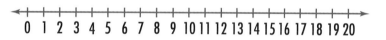

3. $1 \times 6 =$ _____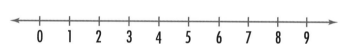

Draw a picture to show the multiplication problem. Write the product.

4. $5 \times 6 =$ _____   5. $6 \times 4 =$ _____

Multiply.

6. $\begin{array}{r} 7 \\ \times\,6 \\ \hline \end{array}$ 7. $\begin{array}{r} 6 \\ \times\,5 \\ \hline \end{array}$ 8. $\begin{array}{r} 1 \\ \times\,6 \\ \hline \end{array}$ 9. $\begin{array}{r} 6 \\ \times\,4 \\ \hline \end{array}$ 10. $\begin{array}{r} 3 \\ \times\,6 \\ \hline \end{array}$

Solve. Remember to label your answers.

11. If each flower has 6 petals, how many petals are on 8 flowers?

_____

12. Mrs. Swift bought 4 packages of muffins. Each package had 6 muffins. How many muffins did she buy altogether?

_____

MIXED Practice

Complete each sentence.

1. In $3 \times 4 = 12$, 3 and 4 are the _____.

2. In $2 \times 7 = 14$, _____ is the product.

3. In $12 + 16 = 28$, 12 and 16 are the _____ and 28 is the _____.

4. $6 + 6 + 6 + 6 = 24$ equals _____ $\times 6 = 24$.

Solve. Remember to label your answers.

5. a. Dina spent $14.99 on a shirt and $10.99 on a scarf. How much did she spend in all? _____

   b. If Dina gave the clerk $30, how much change did she receive? _____

Name _____

Multiplying by 7

Multiply.

1. $0 \times 7 =$ _____ $1 \times 7 =$ _____

   $2 \times 7 =$ _____ $3 \times 7 =$ _____

   $4 \times 7 =$ _____ $5 \times 7 =$ _____

   $6 \times 7 =$ _____ $7 \times 7 =$ _____

   $8 \times 7 =$ _____ $9 \times 7 =$ _____

Multiply the number in the center by each number in the inside ring. Write the products in the outside ring.

2.
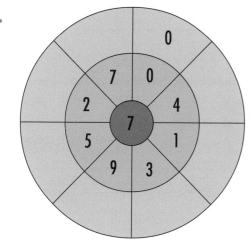

3.
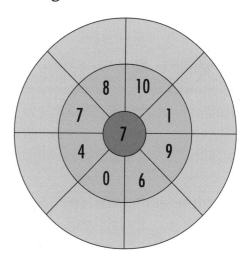

Multiply.

4. $9 \times 7 =$ _____   5. _____ $= 7 \times 5$ 6. $7 \times 8 =$ _____

7. _____ $= 0 \times 7$ 8. $2 \times 7 =$ _____   9. _____ $= 7 \times 3$

10. $\begin{array}{r} 7 \\ \times\, 4 \\ \hline \end{array}$ 11. $\begin{array}{r} 8 \\ \times\, 7 \\ \hline \end{array}$ 12. $\begin{array}{r} 7 \\ \times\, 1 \\ \hline \end{array}$ 13. $\begin{array}{r} 6 \\ \times\, 7 \\ \hline \end{array}$ 14. $\begin{array}{r} 3 \\ \times\, 7 \\ \hline \end{array}$

Solve. Remember to label your answers.

15. 6 weeks = _____ days

16. _____ days = 8 weeks

17. Jerry planted 8 rows of tomato plants with 7 plants in each row. How many tomato plants did Jerry plant in all?

_____

18. A theatre reserved 7 rows of seats for a group of students. There were 9 seats in each row. How many seats were reserved?

_____

MIXED Practice

Compare. Write <, >, or =.

1. 4,931 ◯ 4,929

2. 868 ◯ 1,021

3. 3,633 ◯ 3,633

Complete each pattern.

4. 63, 76, 89, 102, _____, _____, _____

5. 3rd, 6th, 9th, 12th, _____, _____, _____

Complete the chart.

6.

Wagon Wheels

Wheels	4					
Wagons	1	2	3	4	5	6

Name _____

Multiplying by 8

Multiply.

1. $0 \times 8 =$ _____ $1 \times 8 =$ _____

    $2 \times 8 =$ _____ $3 \times 8 =$ _____

    $4 \times 8 =$ _____ $5 \times 8 =$ _____

    $6 \times 8 =$ _____ $7 \times 8 =$ _____

    $8 \times 8 =$ _____ $9 \times 8 =$ _____

2. Find each product. Then below each problem write the letter
    that matches each answer. You will spell a sentence.

0 = T	56 = G	48 = R	64 = M	8 = X	21 = Y	42 = U
72 = E	16 = I	32 = H	40 = S	24 = F	35 = O

$$
\begin{array}{ccccc}
9 & 2 & 7 & 8 & 8 \\
\times 8 & \times 8 & \times 8 & \times 4 & \times 0 \\
\hline
\end{array}
\qquad
\begin{array}{ccccc}
0 & 8 & 8 & 8 & 5 \\
\times 8 & \times 2 & \times 8 & \times 9 & \times 8 \\
\hline
\end{array}
$$

___ ___ ___ ___ ___   ___ ___ ___ ___ ___

$$
\begin{array}{ccccc}
9 & 8 & 8 & 4 & 8 \\
\times 8 & \times 2 & \times 7 & \times 8 & \times 0 \\
\hline
\end{array}
\qquad
\begin{array}{cc}
2 & 8 \\
\times 8 & \times 5 \\
\hline
\end{array}
$$

___ ___ ___ ___ ___   ___ ___

$$
\begin{array}{ccccc}
5 & 8 & 8 & 0 & 3 \\
\times 8 & \times 2 & \times 1 & \times 8 & \times 7 \\
\hline
\end{array}
\qquad
\begin{array}{cccc}
8 & 7 & 6 & 8 \\
\times 3 & \times 5 & \times 7 & \times 6 \\
\hline
\end{array}
$$

___ ___ ___ ___ ___   ___ ___ ___ ___

Multiply.

3. $8 \times 1 = $ _____ 4. _____ $= 7 \times 8$ 5. $9 \times 8 = $ _____

6. _____ $= 8 \times 3$ 7. $4 \times 8 = $ _____ 8. _____ $= 6 \times 8$

Solve. Remember to label your answers.

9. There are 8 crayons in a pack. How many crayons are in 9 packs?

_____

10. Jonah bought 8 packs of trading cards. There are 8 cards in a pack. How many cards did Jonah buy in all?

_____

MIXED Practice

Add or subtract.

1.	1,329 − 571	2.	4,568 + 4,321	3.	2,331 + 679

4.	2,310 + 4,347	5.	9,104 − 3,099	6.	4,630 − 1,422

Order the numbers from greatest to least.

7. 8,706; 8,607; 8,760

_____; _____; _____

8. 154, 164, 145, 146

_____, _____, _____, _____

Name _____

Multiplying by 9

Multiply.

1. $0 \times 9 =$ _____ $1 \times 9 =$ _____

   $2 \times 9 =$ _____ $3 \times 9 =$ _____

   $4 \times 9 =$ _____ $5 \times 9 =$ _____

   $6 \times 9 =$ _____ $7 \times 9 =$ _____

   $8 \times 9 =$ _____ $9 \times 9 =$ _____

2. _____ $= 8 \times 9$ 3. $1 \times 9 =$ _____   4. _____ $= 9 \times 4$

5. $7 \times 9 =$ _____   6. _____ $= 9 \times 6$ 7. $9 \times 8 =$ _____

8. $\begin{array}{r} 9 \\ \times\,9 \\ \hline \end{array}$ 9. $\begin{array}{r} 0 \\ \times\,9 \\ \hline \end{array}$ 10. $\begin{array}{r} 6 \\ \times\,7 \\ \hline \end{array}$ 11. $\begin{array}{r} 9 \\ \times\,1 \\ \hline \end{array}$ 12. $\begin{array}{r} 9 \\ \times\,5 \\ \hline \end{array}$

13. $\begin{array}{r} 3 \\ \times\,9 \\ \hline \end{array}$ 14. $\begin{array}{r} 9 \\ \times\,4 \\ \hline \end{array}$ 15. $\begin{array}{r} 9 \\ \times\,2 \\ \hline \end{array}$ 16. $\begin{array}{r} 8 \\ \times\,7 \\ \hline \end{array}$ 17. $\begin{array}{r} 9 \\ \times\,6 \\ \hline \end{array}$

Solve. Remember to label your answers.

18. There are 9 players on a baseball team. How many baseball players will there be in a tournament of 4 teams? _____

19. There are 2 pints in 1 quart. How many pints are there in 9 quarts? _____

Multiply the number in the center by each number in the inside ring.
Write the products in the outside ring.

20.

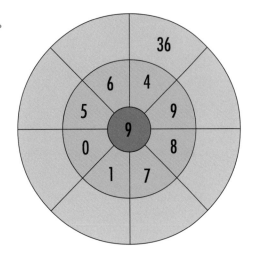

21.

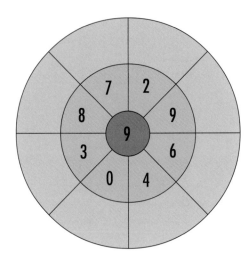

Name _____

Factors of 8 or 9

Find and ring two factors and their product. In each multiplication problem, one of the factors will be 8 or 9. The multiplication problems read left to right or top to bottom.

After you ring each multiplication fact, write it below.

6	7	2	6	5	1	9	9	7	9	8	4
2	9	18	3	3	8	3	2	3	7	9	3
4	63	7	2	8	8	9	18	6	63	6	6
1	5	4	4	5	9	45	5	2	2	9	1
5	3	1	2	0	9	0	2	9	6	54	4
8	9	72	3	7	5	0	1	8	9	1	0
1	27	8	0	4	8	32	6	72	0	9	6
8	3	24	8	5	40	7	6	4	9	36	7
4	8	5	0	8	6	4	8	1	0	3	2
7	24	9	9	8	8	6	48	8	5	0	9

$9 \times 7 = 63$ _____

_____ _____ _____

_____ _____ _____

_____ _____ _____

_____ _____ _____

_____ _____ _____

_____

Name _____

Multiplying by 11 or 12

Multiply.

1. 0 × 11 = ____   0 × 12 = ____

    1 × 11 = ____   1 × 12 = ____

    2 × 11 = ____   2 × 12 = ____

    3 × 11 = ____   3 × 12 = ____

    4 × 11 = ____   4 × 12 = ____

    5 × 11 = ____   5 × 12 = ____

    6 × 11 = ____   6 × 12 = ____

    7 × 11 = ____   7 × 12 = ____

    8 × 11 = ____   8 × 12 = ____

    9 × 11 = ____   9 × 12 = ____

    10 × 11 = ____   10 × 12 = ____

    11 × 11 = ____   11 × 12 = ____

    12 × 11 = ____   12 × 12 = ____

2.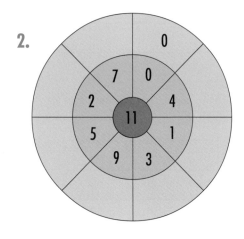

3.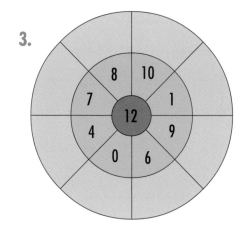

Name _____

Using a Multiplication Table

Multiply. Use the multiplication table in your textbook.

1. _____ $= 6 \times 5$ 2. $8 \times 4 =$ _____   3. _____ $= 1 \times 8$

4. $6 \times 2 =$ _____   5. _____ $= 6 \times 7$ 6. $8 \times 8 =$ _____

7. _____ $= 8 \times 5$ 8. $5 \times 1 =$ _____   9. _____ $= 7 \times 6$

10. $\begin{array}{r} 7 \\ \times\, 9 \\ \hline \end{array}$ 11. $\begin{array}{r} 9 \\ \times\, 6 \\ \hline \end{array}$ 12. $\begin{array}{r} 3 \\ \times\, 9 \\ \hline \end{array}$ 13. $\begin{array}{r} 7 \\ \times\, 7 \\ \hline \end{array}$ 14. $\begin{array}{r} 1 \\ \times\, 4 \\ \hline \end{array}$

15. $\begin{array}{r} 1 \\ \times\, 7 \\ \hline \end{array}$ 16. $\begin{array}{r} 6 \\ \times\, 8 \\ \hline \end{array}$ 17. $\begin{array}{r} 6 \\ \times\, 6 \\ \hline \end{array}$ 18. $\begin{array}{r} 5 \\ \times\, 9 \\ \hline \end{array}$ 19. $\begin{array}{r} 2 \\ \times\, 1 \\ \hline \end{array}$

Write the factors for each product. Use the multiplication table.

20. ___ $\times$ ___ $= 56$ 21. $21 =$ ___ $\times$ ___   22. ___ $\times$ ___ $= 81$

23. $28 =$ ___ $\times$ ___   24. ___ $\times$ ___ $= 77$ 25. $84 =$ ___ $\times$ ___

Solve. Remember to label your answers.

26. Write 3 multiplication facts that have a product of 18.

_____   _____   _____

27. The product is 30. One of the factors is 6. What is the other factor?

_____

Name _____

Finding the Missing Factor

Write each missing factor.

1. $9 \times \underline{\quad} = 18$
2. $63 = \underline{\quad} \times 7$
3. $2 \times \underline{\quad} = 2$

4. $18 = \underline{\quad} \times 1$
5. $9 \times \underline{\quad} = 54$
6. $19 = \underline{\quad} \times 19$

7. $5 \times \underline{\quad} = 15$
8. $14 = \underline{\quad} \times 2$
9. $7 \times \underline{\quad} = 49$

10. $9 = \underline{\quad} \times 3$
11. $\underline{\quad} \times 8 = 40$
12. $60 = \underline{\quad} \times 1$

Solve. Remember to label your answers.

13. Twenty-seven people will be at Curt's birthday party. Curt wants to give party favors to his guests. There are 3 favors in each box. How many boxes must Curt buy so that everyone receives one favor?

    _____

14. Janice would like to read a 72-page book in 8 days. If she reads the same number of pages each day, how many pages will Janice have to read each day?

    _____

15. Mrs. Thomas wants to cut a rectangular cake into 24 pieces. She makes 4 rows. How many pieces does she need to make from each row?

    _____

Name _____

Three or More Factors

Multiply.

1. $(2 \times 4) \times 8 =$ _____

2. $(1 \times 9) \times 7 =$ _____

3. _____ $= 3 \times (2 \times 5)$

4. $(2 \times 4) \times 5 =$ _____

5. $7 \times (4 \times 2) =$ _____

6. _____ $= (2 \times 5) \times 8$

7. $(2 \times 5) \times (3 \times 1) =$ _____

8. $(2 \times 2) \times (2 \times 5) =$ _____

9. _____ $= (4 \times 2) \times (5 \times 0)$

10. _____ $= (6 \times 1) \times (3 \times 2)$

Multiply. Make your own grouping.

11. $3 \times 3 \times 4 =$ _____

12. $3 \times 4 \times 2 =$ _____

Solve. Remember to label your answers.

13. Zack made 3 glasses of lemonade each day. Each glass had 2 slices of lemon. He did this every day for 4 days. How many slices of lemon did he use altogether?

    _____

14. A package contains 4 sheets of stickers. Each sheet has 5 stickers. How many stickers are in 2 packages?

    _____

15. A cookie sheet has 4 rows of cookies. Each row contains 5 cookies. Julian fills two cookie sheets. How many cookies did Julian make?

    _____

Name _____

Multiplying to Find Volume

Volume is the number of cubic units something holds.

To find the volume of these figures, multiply length × width × height.

You also could think: layers × rows × columns

1.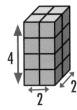
length = 2 units
width = 2 units
height = 4 units

_____ × _____ × _____

= _____ cubic units

2.
length = ___ units
width = ___ units
height = ___ units

_____ × _____ × _____

= _____ cubic units

3.

_____ × _____ × _____

= _____ cubic units

4.

_____ × _____ × _____

= _____ cubic units

Solve. Remember to label your answers.

5. Mrs. Simons is shopping for a freezer. Freezer A has a volume of 9 cubic feet. The volume of freezer B is 12 cubic feet. Which freezer holds more food?

_____

6. Find the volume.

9 × 2 × 4 = _____ cubic units

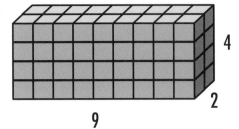

Name _____

Problem-Solving Skill: Combinations

Complete the table.

1.

Papa Joe's Pizzas

Number of Pizza Toppings	Available Sizes: Small, Medium, Large, Extra Large	Number of Possible Pizza Combinations
1	4	
2	4	
3	4	
4	4	

Use the menu to answer the questions.

2. You want 2 pizzas, each with one topping. How many possible combinations do you have with the toppings at Papa Joe's Pizzeria?

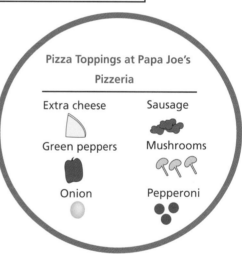

Pizza Toppings at Papa Joe's Pizzeria

Extra cheese   Sausage

Green peppers   Mushrooms

Onion   Pepperoni

_____

3. Describe how you found your answer.

_____

_____

4. Suppose you want 3 pizzas, each with 1 topping. How many possible combinations would you have? _____

Name _____

Chapter 7 Review

Write a multiplication equation for each picture.

1.

2.

3.

_____   _____   _____

Multiply.

4. _____ $= 2 \times 9$ 5. $6 \times 7 =$ _____   6. _____ $= 5 \times 8$

7. $4 \times 7 =$ _____   8. _____ $= 7 \times 3$ 9. $4 \times 9 =$ _____

10. _____ $= 7 \times 8$ 11. $9 \times 9 =$ _____   12. _____ $= 9 \times 7$

13. $\begin{array}{r} 6 \\ \times 3 \\ \hline \end{array}$ 14. $\begin{array}{r} 0 \\ \times 8 \\ \hline \end{array}$ 15. $\begin{array}{r} 6 \\ \times 9 \\ \hline \end{array}$ 16. $\begin{array}{r} 8 \\ \times 6 \\ \hline \end{array}$ 17. $\begin{array}{r} 8 \\ \times 4 \\ \hline \end{array}$

18. $\begin{array}{r} 2 \\ \times 8 \\ \hline \end{array}$ 19. $\begin{array}{r} 4 \\ \times 6 \\ \hline \end{array}$ 20. $\begin{array}{r} 3 \\ \times 9 \\ \hline \end{array}$ 21. $\begin{array}{r} 6 \\ \times 5 \\ \hline \end{array}$ 22. $\begin{array}{r} 7 \\ \times 5 \\ \hline \end{array}$

Multiply.

23. $(2 \times 1) \times 7 =$ _____   24. _____ $= (2 \times 2) \times 9$

25. _____ $= 5 \times (4 \times 2)$ 26. $(6 \times 1) \times (9 \times 0) =$ _____

Write each missing factor.

27. $7 \times$ _____ $= 63$ 28. _____ $\times 8 = 64$ 29. $7 \times$ _____ $= 21$

© Calvert School

Solve. Remember to label your answers.

30. There are 6 cans of soup in a package. How many cans are in 6 packages?

_____

31. Use the Camp Schedule to find out how many possible activity combinations there are for one day of camp.

_____

_____

Summer Camp Schedule

Pick 1 camp activity for each morning and each afternoon:

Morning	Afternoon
Swimming	Crafts
Biking	Building
Canoeing	Movies

32. There are 8 canoes at a dock. Each canoe has 2 paddles. How many paddles are there in all?

_____

_____

33. Explain how you can check a multiplication problem using a multiplication table.

_____

_____

Name _____

Chapter 7 Test Prep

Ring the letter of the correct answer.

1. Which array shows 6 × 6?

   a.

   b. ✳ ✳ ✳ ✳ ✳ ✳
   ✳ ✳ ✳ ✳ ✳ ✳
   ✳ ✳ ✳ ✳ ✳ ✳
   ✳ ✳ ✳ ✳ ✳ ✳
   ✳ ✳ ✳ ✳ ✳ ✳
   ✳ ✳ ✳ ✳ ✳ ✳

   c. ●●●●●●
   ●●●●●●

   d. none of these

2. Joe has 3 shirts—blue, red, and white. He has 4 pairs of shorts—red, green, yellow, and blue. How many different shirt-and-shorts outfits does he have?

   a. 3 outfits   b. 6 outfits   c. 9 outfits   d. 12 outfits

3. There are 56 cookies at a bakery. There are 8 different types of cookies. There are the same number of each type of cookie. How many of each type of cookie are there?

   a. 6 cookies   b. 7 cookies   c. 8 cookies   d. 9 cookies

4. To solve 8 × 2 × 3, you can multiply (8 × 2) × 3, or you can multiply 8 × (2 × 3). This is called using the _____.

   a. Associative Property of Multiplication

   b. Zero Property of Multiplication

   c. Commutative Property of Multiplication

   d. Identity Property of Multiplication

5. $9 \times 8 =$ _____

    a. 70   b. 71   c. 72   d. 73

6. $8 \times 6 =$ _____

    a. 48   b. 49   c. 56   d. 58

7. $9 \times 9 =$ _____

    a. 72   b. 74   c. 80   d. 81

8. $7 \times 7 =$ _____

    a. 48   b. 49   c. 56   d. 63

9. $3 \times 2 \times 7 =$ _____

    a. 40   b. 42   c. 45   d. 48

10. There were 8 dogs in a dog park. Altogether, how many dog paws were there?

    a. 28 paws   b. 32 paws   c. 36 paws   d. 40 paws

Name _____

Relating Multiplication and Division

Complete each fact family.

1. $7 \times 9 =$ _____

   $9 \times 7 =$ _____

   $63 \div 9 =$ _____

   $63 \div 7 =$ _____

2. $7 \times 4 =$ _____

   $4 \times 7 =$ _____

   $28 \div 7 =$ _____

   $28 \div 4 =$ _____

Use each group of numbers to write a fact family.

3. 4, 9, 36

   _____ _____

   _____ _____

4. 7, 8, 56

   _____ _____

   _____ _____

5. 6, 9, 54

   _____ _____

   _____ _____

6. 5, 9, 45

   _____ _____

   _____ _____

7. 3, 6, 18

   _____ _____

   _____ _____

8. 6, 6, 36

   _____ _____

Complete each equation.

9. $3 \times \boxed{} = 27$

   $27 \div 3 = \boxed{}$

10. $2 \times \boxed{} = 14$

    $14 \div 2 = \boxed{}$

11. $7 \times \boxed{} = 21$

    $21 \div 7 = \boxed{}$

Ring the equation that is *not* part of the fact family.

12. $5 \times 4 = 20$

$20 \div 4 = 5$

$24 \div 4 = 6$

$4 \times 5 = 20$

$20 \div 5 = 4$

13. $10 \times 3 = 30$

$6 \times 5 = 30$

$30 \div 6 = 5$

$30 \div 5 = 6$

$5 \times 6 = 30$

14. $18 \div 9 = 2$

$9 + 2 = 11$

$18 \div 2 = 9$

$2 \times 9 = 18$

$9 \times 2 = 18$

Solve.

15. Write the fact family for the factors 5 and 7.

_____

16. Explain why there are only two equations for the fact family 6, 6, and 36.

_____

_____

MIXED Practice

Multiply.

1. $\begin{array}{r} 9 \\ \times 3 \\ \hline \end{array}$

2. $\begin{array}{r} 6 \\ \times 7 \\ \hline \end{array}$

3. $\begin{array}{r} 8 \\ \times 6 \\ \hline \end{array}$

4. $\begin{array}{r} 5 \\ \times 5 \\ \hline \end{array}$

5. $\begin{array}{r} 7 \\ \times 4 \\ \hline \end{array}$

6. $\begin{array}{r} 7 \\ \times 7 \\ \hline \end{array}$

7. $\begin{array}{r} 7 \\ \times 9 \\ \hline \end{array}$

8. $\begin{array}{r} 8 \\ \times 8 \\ \hline \end{array}$

Name _____

Division as Repeated Subtraction

Division is repeated subtraction.

Division:	**Repeated subtraction:**

$12 \div 4 =$ _____

THINK How many 4s are in 12?

You can subtract 4 three times.

So, $12 \div 4 = 3$.

$$\begin{array}{r} 12 \\ -\ 4 \\ \hline 8 \\ -\ 4 \\ \hline 4 \\ -\ 4 \\ \hline 0 \end{array}$$ ← subtract 4
← subtract 4
← subtract 4

Divide. Use repeated subtraction. Show your work.

1. $8 \div 2 =$ _____ $\begin{array}{r} 8 \\ -\ 2 \\ \hline \end{array}$

2. $15 \div 5 =$ _____ $\begin{array}{r} 15 \\ -\ 5 \\ \hline \end{array}$

3. $27 \div 9 =$ _____ $\begin{array}{r} 27 \\ -\ 9 \\ \hline \end{array}$

4. $16 \div 8 =$ _____ $\begin{array}{r} 16 \\ -\ 8 \\ \hline \end{array}$

5. $21 \div 7 =$ _____ $\begin{array}{r} 21 \\ -\ 7 \\ \hline \end{array}$

6. $30 \div 6 =$ _____ $\begin{array}{r} 30 \\ -\ 6 \\ \hline \end{array}$

Name _____

Dividing by 2 or 5

1. Use the words dividend, quotient, and divisor to label the parts of a division equation.

$$12 \div 2 = 6$$

Write a division equation for each picture.

2.

_____

3.

_____

4.

_____

Divide. Make groups of 2.

5. 18 counters

   $18 \div 2 =$ _____

6. 12 counters

   $12 \div 2 =$ _____

7. 14 counters

   $14 \div 2 =$ _____

Divide. Make groups of 5.

8. 25 counters

   $25 \div 5 =$ _____

9. 45 counters

   $45 \div 5 =$ _____

10. 30 counters

    $30 \div 5 =$ _____

Divide.

11. $16 \div 2 =$ _____

12. $10 \div 2 =$ _____

13. $20 \div 5 =$ _____

14. $2\overline{)8}$

15. $5\overline{)20}$

16. $5\overline{)15}$

17. $5\overline{)25}$

Solve. Remember to label your answers.

18. A librarian displays 40 books. She puts 5 books on each shelf. How many shelves does she use?

_____

19. David has 30 dimes that he puts in piles of 5. How many piles does he make?

_____

MIXED Practice

Add or subtract.

1. $\begin{array}{r} 56 \\ + 18 \\ \hline \end{array}$

2. $\begin{array}{r} 49 \\ - 21 \\ \hline \end{array}$

3. $\begin{array}{r} 122 \\ + \ 88 \\ \hline \end{array}$

4. $\begin{array}{r} 17,651 \\ - \ 5,372 \\ \hline \end{array}$

5. $\begin{array}{r} 6,914 \\ + 2,231 \\ \hline \end{array}$

6. $\begin{array}{r} 5,000 \\ - 3,219 \\ \hline \end{array}$

Write the place value of the 6 in each number.

7. 61,403

8. 15,621

9. 1,416

_____   _____   _____

Write the amount of money.

10. two one-dollar bills, six quarters, one dime, fifteen pennies

_____

11. three quarters, ten nickels, six pennies _____

166 Grade 3 • Chapter 8

Name _____

Dividing by 3 or 4

Draw lines to make equal groups. Complete the division problem.

1. Make groups of 3.

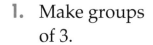

    $15 \div 3 =$ _____

2. Make 4 equal groups.

    $12 \div 4 =$ _____

3. Make 3 equal groups.

    $9 \div 3 =$ _____

Divide.

4. 12 counters
   3 equal groups

   _____ in each group

   $12 \div 3 =$ _____

5. 18 counters
   3 equal groups

   _____ in each group

   _____ $= 18 \div 3$

6. 24 counters
   4 in each group

   _____ equal groups

   $4\overline{)24}$

7. 21 counters
   3 in each group
   _____ equal groups

   $3\overline{)21}$

8. 20 counters
   4 equal groups
   _____ in each group

   $20 \div 4 =$ _____

9. 16 counters
   4 in each group
   _____ equal groups

   _____ $= 16 \div 4$

10. 30 counters
    3 equal groups
    _____ in each group

    _____ $= 30 \div 3$

11. 28 counters
    4 in each group
    _____ equal groups

    $28 \div 4 =$ _____

12. 27 counters
    3 equal groups
    _____ in each group

    $3\overline{)27}$

Solve. Remember to label your answers.

13. Jeremy planted 28 seeds in 4 equal rows. How many seeds did he plant in each row?

14. Becky displays her 24 stuffed animals equally on 3 shelves. How many stuffed animals are on each shelf?

MIXED Practice

Write the missing factors.

1. $5 \times \boxed{} = 40$

2. $81 = \boxed{} \times 9$

3. $28 = \boxed{} \times 4$

4. $6 \times \boxed{} = 36$

Complete each pattern.

5. 21, 25, 29, 33, ____, ____, ____

6. 16, 10, 15, 9, ____, ____, ____

Solve. Remember to label your answers.

7. The Bagel Shop sold 856 bagels on Monday and 715 on Tuesday. How many bagels were sold altogether?

8. Aaron wants to buy a backpack that costs $24.98. He has already saved $16.49. How much more money does he need?

© Calvert School

Name _____

Problem-Solving Skill: Choosing an Operation

Solve. Remember to label your answers.

Maggie bought 4 packs of invitations. Each pack had 8 invitations. How many invitations did Maggie buy?

1. What do you need to find?

   _____

2. What facts do you know?

   _____

3. Are you combining, taking away, putting together equal groups, or separating items equally?

   _____

4. Should you add, subtract, multiply, or divide?

   _____

5. What is the answer?

   _____

6. Check the answer.

   _____

   _____

Choose an operation. Then solve. Remember to label your answers.

7. Renee had 25 guppies in her aquarium. A month later, 17 more guppies were born. How many guppies are there in all?

   _____

8. A taxi company ordered 36 new tires. Each taxi needs 4 new tires. How many taxis can get new tires?

   _____

9. Tickets to the Jazz Festival cost $8 each. How much do 5 tickets cost? _____

Ring the equation that solves the problem.

10. Alice has $25. She spends $24 on stuffed animals. How much change does she receive?
    a. $25 − 11 = $14
    b. $25 − 5 = $20
    c. $25 − 24 = $1

11. Tina's cat had 8 kittens in 2 litters. There were the same number of kittens in each litter. How many kittens were in each litter?
    a. 8 − 2 = 6
    b. 8 × 2 = 16
    c. 8 ÷ 2 = 4

Write the equation that solves the problem.

12. A farmer had 36 chickens. He sold 6 chickens. How many chickens are left?

    _____

13. Bonnie displayed 12 photos equally on 6 pages. How many photos are on each page?

    _____

MIXED Practice

Multiply.

1. $\begin{array}{r} 8 \\ \times\, 4 \\ \hline \end{array}$

2. $\begin{array}{r} 3 \\ \times\, 3 \\ \hline \end{array}$

3. $\begin{array}{r} 9 \\ \times\, 6 \\ \hline \end{array}$

4. $\begin{array}{r} 7 \\ \times\, 5 \\ \hline \end{array}$

Name _____

Dividing with 0 and 1

Divide.

1. $6 \div 1 =$ _____

2. _____ $= 9 \div 1$

3. $0 \div 7 =$ _____

4. _____ $= 8 \div 1$

5. $7 \div 7 =$ _____

6. _____ $= 6 \div 6$

7. $0 \div 3 =$ _____

8. _____ $= 3 \div 3$

9. $0 \div 5 =$ _____

10. $1\overline{)5}$

11. $9\overline{)0}$

12. $4\overline{)4}$

13. $3\overline{)0}$

Write the sign to make the problem correct. Use $\times$ or $\div$.

Hint: Some problems may have more than one correct answer.

14. $9 \bigcirc 9 = 1$

15. $4 = 4 \bigcirc 1$

16. $2 \bigcirc 2 = 1$

17. $3 \bigcirc 1 = 3$

18. $0 = 0 \bigcirc 3$

19. $4 \bigcirc 4 = 1$

Solve. Remember to label your answers.

20. A florist had 6 flowers to plant in 6 pots. If each pot gets the same number of flowers, how many flowers will be in each pot?

_____

21. The florist did not have any roses. Four customers wanted to buy roses. How many roses did each customer buy?

_____

Name _____

Dividing by 6 or 7

Complete each division equation.

1. How many groups of 6?

    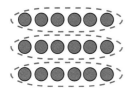

    18 ÷ 6 = _____

2. How many groups of 7?

    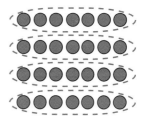

    28 ÷ 7 = _____

Divide. Make groups of 6.

3. 24 counters

    24 ÷ 6 = _____

4. 12 counters

    12 ÷ 6 = _____

5. 36 counters

    36 ÷ 6 = _____

Divide. Make groups of 7.

6. 35 counters

    35 ÷ 7 = _____

7. 7 counters

    7 ÷ 7 = _____

8. 21 counters

    21 ÷ 7 = _____

Divide.

9. $6\overline{)42}$ 10. $7\overline{)35}$ 11. $6\overline{)48}$ 12. $7\overline{)14}$ 13. $7\overline{)49}$

14. Use the words dividend, quotient, and divisor. Label the parts of the division equation.

    56   ÷   7   =   8

    ↑   ↑   ↑

    _____   _____   _____

Divide. Use counters.

15. 42 counters
    7 equal groups
    _____ in each group

    42 ÷ 7 = _____

16. 42 counters
    6 in each group
    _____ equal groups

    42 ÷ 6 = _____

17. 36 counters
    6 equal groups
    _____ in each group

    $6\overline{)36}$

Solve. Remember to label your answers.

18. Draw a picture that shows
    28 ÷ 7 = 4.

19. Draw a picture that shows
    30 ÷ 6 = 5.

20. Caitlin's grandparents took a trip around the world. They were gone for 63 days. How many weeks were they away?

    _____

21. There are 48 balloons in 6 packages. Each package contains the same number of balloons. How many balloons are in each package?

    _____

Name _____

Dividing by 8, 9, 10, 11, or 12

Draw lines to show the number of equal groups.
Complete the equation.

1. Make 9 equal groups.

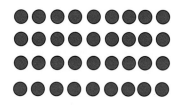

$36 \div 9 =$ _____

2. Make 8 equal groups.

$8\overline{)24}$

3. Draw a picture that shows
$20 \div 10 = 2$.

4. Draw a picture that shows
$22 \div 11 = 2$.

Divide. Make equal groups.

5. 24 counters

$24 \div 8 =$ _____

$8\overline{)24}$

6. 27 counters

$27 \div 9 =$ _____

$9\overline{)27}$

7. 36 counters

$36 \div 12 =$ _____

$12\overline{)36}$

Divide.

8. $72 \div 9 =$ _____   9. $36 \div 9 =$ _____   10. $96 \div 8 =$ _____

11. $63 \div 9 =$ _____   12. $48 \div 8 =$ _____   13. $88 \div 8 =$ _____

14. $9 \div 9 =$ _____   15. $16 \div 8 =$ _____   16. $90 \div 10 =$ _____

17. $8\overline{)56}$ 18. $9\overline{)81}$ 19. $8\overline{)64}$ 20. $9\overline{)45}$ 21. $8\overline{)72}$

MIXED Practice

Compare. Write <, >, or =.

1. $1{,}765 \bigcirc 1{,}675$ 2. $630 \bigcirc 645$ 3. $82{,}261 \bigcirc 82{,}621$

4. $8 + 4 \bigcirc 9 + 4$ 5. $4 + 4 + 4 \bigcirc 3 + 3 + 3 + 3$

Solve. Remember to label your answers.

6. Samantha started practicing the violin at 12:25 P.M. and stopped practicing at 1:30 P.M. How long did she practice?

_____

7. There are 8 hot dogs in a package. How many packages are needed for 40 people to each have 1 hot dog?

_____

8. Each box holds 6 rows of soup cans with 4 cans in each row. How many soup cans are in 3 boxes?

_____

Name _____

Equal and Not Equal

An equation shows that two quantities are equal. Use the symbol =.
It is read as *is equal to*.

$$7 + 2 = 2 + 7 \qquad 6 \times 4 = 4 \times 6 \qquad 3 \times (2 + 5) = 3 \times 7$$

When two quantities are not equal, use the symbol ≠. It is read
as *is not equal to*.

$$5 + 4 \neq 2 + 1 \qquad 7 \times 2 \neq 7 + 2 \qquad 15 - 7 \neq 2 + 10$$

To decide which sign, = or ≠, is correct, first calculate the math on each side.

$$3 \times 4 \bigcirc 7 + 8$$

$$3 \times 4 = 12 \qquad 7 + 8 = 15$$

$$12 < 15$$

$$3 \times 4 \,\cancel{=}\, 7 + 8$$

Calculate the math.
Compare the amounts.
The amounts are not equal.
Write ≠.

Compare. Write = or ≠.

1. $16 - 8 \bigcirc 12 + 4$ 2. $10 + 3 \bigcirc 12 + 1$ 3. $9 \times 4 \bigcirc 30 + 5$

4. $18 \div 6 \bigcirc 4 \times 1$ 5. $6 \times 6 \bigcirc 40 - 4$ 6. $25 \div 5 \bigcirc 0 \times 5$

7. $63 \div 9 \bigcirc 63 \div 7$ 8. $2 \times 7 \bigcirc 4 \times 4$ 9. $20 + 20 \bigcirc 8 \times 6$

10. $(4 \times 2) + 1 \bigcirc (4 \times 3) - 1$ 11. $(21 \div 3) + 4 \bigcirc 3 + (4 \times 2)$

    Hint: Calculate what is inside the parentheses first.

12. Write a number in the blank to make the equation true.

    $$4 + (3 \times \underline{\hspace{1cm}}) = 1 + (3 \times 3)$$

Name _____

Remainders

1. Use the words remainder, quotient, dividend, and divisor
to label the parts of the division problem.

$$\begin{array}{r} 5\ \text{R2} \\ 3\overline{)17} \\ -15 \\ \hline 2 \end{array}$$

    (3) _____   (2) _____

    (17) _____   (5 R2) _____

Divide. Use an R to show each remainder.

2. $5\overline{)16}$ 3. $7\overline{)18}$ 4. $3\overline{)10}$ 5. $8\overline{)27}$

6. $5\overline{)24}$ 7. $3\overline{)19}$ 8. $4\overline{)22}$ 9. $6\overline{)41}$

10. $7\overline{)27}$ 11. $9\overline{)85}$ 12. $8\overline{)47}$ 13. $3\overline{)29}$

Solve. Remember to label your answers.

14. George has 37 pennies. He makes stacks of 5 pennies each. How many stacks can he make? How many pennies will be left over?

_____

15. Sammy has 13 black socks in his drawer. How many pairs does he have? Are there any extra socks? If yes, how many?

_____

© Calvert School

Name _____

Problem-Solving Application: Interpreting Remainders

Think about:
- the meaning of the remainder.
- what the question asks.
- if it make sense to drop the remainder.
- if the quotient has to be increased by one.

Solve. Remember to label your answers.

1. A troop of 15 scouts is going to the zoo. The zoo's rules say that there must be 1 adult with every 4 children. What is the least number of adults needed?

2. The tables at a restaurant each seat 4 people. If 4 people are seated at each table, how many tables are needed to seat 27 people?

3. A baker packs 6 donuts in each bag. How many bags can he fill if he has 53 donuts?

4. Andrea took $10 to the zoo to buy souvenirs. What is the greatest number she can buy if each souvenir costs $3?

MIXED Practice

Solve.

1. Which numbers should you switch so each sum is 13?

$$
\begin{array}{ccc}
2 & 5 & 8 \\
6 & 3 & 4 \\
+\,5 & +\,1 & +\,5 \\
\hline
\end{array}
$$

Name _____

Checking Division by Multiplying

Multiplication and division are opposite operations.
Multiplication can be used to check the answer to a division problem.

$$\begin{array}{r} 6\text{ R}3 \\ 4\overline{)27} \\ -24 \\ \hline 3 \end{array}$$

6 groups of 4 = 24

Remainder

$$\begin{array}{r} 6 \\ \times\ 4 \\ \hline 24 \\ +\ 3 \\ \hline 27 \end{array}$$

Put the 6 groups of 4 back together.

Add the remainder.
The final sum is equal to the dividend, so the quotient is correct.

If the dividend and the final answer in the check are not the same, solve the division problem again.

More Examples

A.   Check

$$\begin{array}{r} 6\text{ R}2 \\ 7\overline{)44} \\ -42 \\ \hline 2 \end{array}$$

equal

$$\begin{array}{r} 6 \\ \times\ 7 \\ \hline 42 \\ +\ 2 \\ \hline 44 \end{array}$$

B.   Check

$$\begin{array}{r} 8\text{ R}5 \\ 6\overline{)53} \\ -48 \\ \hline 5 \end{array}$$

equal

$$\begin{array}{r} 8 \\ \times\ 6 \\ \hline 48 \\ +\ 5 \\ \hline 53 \end{array}$$

Divide. Then check.

1. $6\overline{)49}$ Check

2. $3\overline{)26}$ Check

3. $6\overline{)51}$ Check

4. $8\overline{)45}$ Check

5. $4\overline{)19}$ Check

6. $9\overline{)34}$ Check

Name _____

Chapter 8 Review

Write the letter of the correct answer.

____ 1. the answer to a division problem

____ 2. the 4 in 12 ÷ 4

____ 3. the 35 in 7)$\overline{35}$

____ 4. the amount sometimes left over in a division problem

____ 5. the answer when a number is divided by itself

a. divisor

b. remainder

c. one

d. quotient

e. dividend

Use each group of numbers to write a fact family.

6. 7, 9, 63

7. 6, 8, 48

Complete each equation.

8. 14 counters
   2 equal groups
   ☐ in each group
   $14 ÷ 2 = $ ☐   2)$\overline{14}$

9. 24 counters
   3 in each group
   ☐ equal groups
   $24 ÷ 3 = $ ☐   3)$\overline{24}$

10. 21 counters
    3 equal groups
    ☐ in each group
    $21 ÷ 3 = $ ☐   3)$\overline{21}$

11. 25 counters
    5 in each group
    ☐ equal groups
    $25 ÷ 5 = $ ☐   5)$\overline{25}$

Divide. Use an R to show each remainder.

12. $6\overline{)20}$ 13. $8\overline{)27}$ 14. $3\overline{)16}$ 15. $4\overline{)19}$

16. $4\overline{)15}$ 17. $9\overline{)29}$ 18. $5\overline{)26}$ 19. $7\overline{)44}$

20. $2\overline{)19}$ 21. $6\overline{)27}$ 22. $7\overline{)18}$ 23. $9\overline{)48}$

Solve. Remember to label your answers.

24. How many $5 books can be purchased with $21?

_____

25. There are 25 people waiting to get on a ride at the amusement park. Each car on the ride holds 3 people. What is the least number of cars needed to hold all 25 people?

_____

Name _____

Chapter 8 Test Prep

Ring the letter of the correct answer.

1. $42 \div 6 =$ ____

    a. 5   b. 5 R3   c. 6 R3   d. 7

2. $64 \div 8 =$ ____

    a. 6   b. 6 R4   c. 7 R6   d. 8

3. $72 \div 9 =$ ____

    a. 6   b. 7   c. 8   d. 9

4. $6\overline{)45}$

    a. 7   b. 7 R3   c. 8   d. 8 R3

5. $6\overline{)56}$

    a. 7 R4   b. 8   c. 8 R2   d. 9 R2

6. There are 26 students going to the zoo. Three students can ride in each car. What is the least number of cars needed to transport all the students?

    a. 8   b. 8 R2   c. 9   d. none of these

7. Mitzi had 4 dimes and 8 pennies. What is the value of all the coins?
   a. $(4 \times 10) + 8 = 48¢$
   b. $4 + 10 + 8 = 22¢$
   c. $48 + 10 = 58¢$

8. Perry watched 13 birds at a feeder on Saturday. On Sunday, he watched 27 birds. How many birds did he watch on the two days?
   a. $13 + 27 = 40$
   b. $27 - 13 = 14$
   c. $2 \times 14 = 28$

9. Elise counted 16 octopus tentacles. Each octopus has 8 tentacles. How many octopuses did she see?
   a. $16 \div 8 = 2$
   b. $16 - 8 = 8$
   c. $16 + 8 = 24$

10. Mr. Johnson printed 56 flyers about a yardsale. Each volunteer took 8 flyers to hang up around town. How many volunteers were there?
    a. $56 - 8 = 48$
    b. $56 + 8 = 64$
    c. $56 \div 8 = 7$

Name _____

Estimating and Measuring to the Nearest Inch, Half-Inch

Estimate each measurement. Then measure each object to the nearest inch or half-inch. Compare. Is your estimate an overestimate or an underestimate?

	Estimate	Actual	Overestimate or Underestimate?
1. height of a magazine	_____	_____	_____
2. length of a stapler	_____	_____	_____
3. length of your foot	_____	_____	_____
4. length of a crayon	_____	_____	_____
5. length of a pair of scissors	_____	_____	_____

Draw a line that has the given length. Use an inch ruler.

6. $5\frac{1}{2}$ in.

7. $1\frac{1}{2}$ in.

8. 4 in.

9. 5 in.

10. $2\frac{1}{2}$ in.

11. $4\frac{1}{2}$ in.

Name _____

Other Customary Units of Length

Refer to the benchmarks in Section 9.2 of your math book as needed.
Ring the best measurement.

1. distance between 2 cities   **a.** 100 in.   **b.** 100 ft   **c.** 100 mi

2. height of a coffee table   **a.** 2 in.   **b.** 2 ft   **c.** 2 yd

3. length of a car   **a.** 15 in.   **b.** 15 ft   **c.** 15 yd

4. height of a refrigerator   **a.** 6 ft   **b.** 6 yd   **c.** 6 mi

5. height of an adult male   **a.** $6\frac{1}{2}$ in.   **b.** $6\frac{1}{2}$ ft   **c.** $6\frac{1}{2}$ yd

Complete.

6. 1 yd = _____ ft   7. 5 yd = _____ ft   8. 1 ft = _____ in.

9. 24 in. = _____ ft   10. 1 yd = ___ in.   11. 72 in. = ____ yd

MIXED Practice

Write each missing factor.

1. $48 = \boxed{} \times 8$ 2. $9 \times \boxed{} = 36$ 3. $25 = \boxed{} \times 5$

Find the amount of change.

	Cost	Money Given	Amount of Change
4.			 ¢
5.			_____ ¢

Name _____

Customary Units of Capacity

Complete.

1. 1 quart = _____ pints

2. _____ cups = 1 pint

3. _____ cups = 1 quart

4. 1 gallon = _____ quarts

Write the abbreviation for each unit of measure.

5. cup _____

6. pint _____

7. quart _____

8. gallon _____

Ring the greater amount.

9. 1 c or 1 pt

10. 1 qt or 1 pt

11. 2 qt or 1 gal

12. 2 qt or 2 c

13. 1 qt or 8 c

14. 3 c or 1 qt

Choose the best measurement. Write a or b.

___ 15. milk in a baby's bottle   a. 1 c   b. 1 gal

___ 16. juice in a juice box   a. 1 c   b. 1 qt

___ 17. full tank of gas in a car   a. 13 pt   b. 13 gal

___ 18. water in a small fish bowl   a. 4 gal   b. 4 c

___ 19. water in an inflatable pool   a. 15 pt   b. 15 gal

___ 20. water in a tea cup   a. 1 c   b. 1 gal

Name _____

Customary Units of Weight

Complete.

1. 1 pound = _____ ounces
2. 1 ton = _____ pounds

Write the abbreviation for each unit of measure.

3. pound = _____
4. ounce = _____
5. ton = _____

Choose the best measurement. Write *a* or *b*.

____ 6. apple   a. ounce   b. pound

____ 7. toy airplane   a. ounce   b. ton

____ 8. real airplane   a. ounce   b. ton

____ 9. pig   a. ounce   b. pound

____ 10. man   a. pound   b. ton

Does each object weigh more or less than a pound? Write *more* or *less*.

11. dinner fork _____
12. television set _____
13. large cat _____
14. piece of paper _____
15. ruler _____
16. chair _____

Solve. Remember to label your answers.

17. Angela bought a box of cookies that weighs 22 ounces. How much more than a pound do the cookies weigh?

_____

18. How many pounds does a 3-ton truck weigh?

_____

Name _____

Customary Unit Fun Facts

Use your knowledge of customary units to ring the best answer for each of the following. Then write the letters you ringed in the corresponding spaces below to name the state that answers the question.

1. The bee hummingbird is the smallest bird in the world and is about _____ long.
   a. 2 inches
   b. 2 feet
   c. 20 inches
   d. 20 feet

2. The standard distance all the way around the bases in baseball is _____.
   k. 120 feet
   l. 120 yards
   m. 120 miles
   n. 120 inches

3. The Washington Monument is just over _____ tall.
   a. 555 feet
   b. 555 inches
   c. $55\frac{1}{2}$ feet
   d. $55\frac{1}{2}$ inches

4. The distance across the widest part of a music CD is about _____ wide.
   p. 1 yard
   q. 3 feet
   r. $2\frac{1}{2}$ inches
   s. $4\frac{1}{2}$ inches

5. Maryland's Chesapeake Bay Bridge is about ___ long.
   k. 4 miles
   l. 150 feet
   m. 99 yards
   n. 1,000 inches

6. The Mississippi River is _____ long.
   a. 2,348 miles
   b. 2,348 yards
   c. 2,348 feet
   d. 2,348 inches

What state in the United States of America has the longest coastline, measuring 6,640 miles?

1	2	3	4	5	6

Name _____

Problem-Solving Skill: Making a Table or Chart

Solve. Remember to label your answers.

1. Tina and Curt are planning to make banners to advertise a fair. Every 3 banners take 7 yards of material. How much material will it take to make 18 banners?

Number of Banners	3	6	9	12		
Yards of Material	7	14				

_____

2. It will take Tina and Curt 2 hours to make 3 banners. How long will it take them to make all 18 banners?

Number of Banners	3	6	9	12		
Number of Hours	2	4				

_____

3. Ice cream will be sold at the fair. One gallon will serve 12 people. How many people can 5 gallons of ice cream serve?

Number of Gallons	1	2	3		
Number of People	12	24			

_____

Name _____

Temperature: Celsius and Fahrenheit

Write each temperature.

1. thirty-five degrees Celsius _____

2. one hundred degrees Fahrenheit _____

Use the thermometers in Section 9.6 of your math textbook to complete the chart.

	What Happens	Celsius Temperature	Fahrenheit Temperature
3.	Water freezes	°C	°F
4.	Water boils	°C	°F
5.	Room temperature		
6.	Normal body temperature		

Ring the best temperature.

7. The temperature outside was ____ °F, so we went swimming.

   a. 55   b. 85

8. Tomorrow the temperature will be ____ °C, so I will wear my hat and gloves.

   a. 5   b. 30

MIXED Practice

Solve. Remember to label your answers.

1. Matt bought a ticket for $7.95. He paid with a twenty-dollar bill. How much change did he receive?

2. There are 15 cats but only 6 cat treats. How many more treats are needed to give each cat 1 treat?

Name _____

Temperature Conversion.

Using the Celsius thermometer, water freezes at 0°C.
Water boils at 100°C.

Using the Fahrenheit thermometer, water freezes
at 32°F. Water boils at 212°F.

**Complete the tables below. Use your calculator
to *convert*, or rename, from one scale to the other.**

To convert from Fahrenheit to Celsius…
• Start with the Fahrenheit temperature.
• Subtract 32.
• Then use your calculator to divide by 1.8.

Temperature Fahrenheit	32°F	50°F	77°F	95°F	104°F	212°F
Temperature Celsius	___°C	___°C	___°C	___°C	___°C	___°C

To convert from Celsius to Fahrenheit…
• Start with the Celsius temperature.
• Use your calculator to multiply by 1.8.
• Then add 32.

Temperature Celsius	0°C	15°C	30°C	45°C	60°C	80°C	100°C
Temperature Fahrenheit	___°F	___°F	___°F	___°F	___°F	___°F	___°F

Name _____

Estimating and Measuring to the Nearest Centimeter, Half-Centimeter

Estimate each measurement. Then measure each object to the nearest centimeter or half-centimeter.

		Estimate	Actual
1.	length of a paper clip	_____	_____
2.	length of a colored pencil	_____	_____
3.	length of the bottom of your shoe	_____	_____
4.	width of your hand (thumb to pinky)	_____	_____
5.	width of your folder	_____	_____

Draw a line that has the given length. Use a metric ruler.

6. 13 cm

7. 3 cm

8. $8\frac{1}{2}$ cm

9. 1 cm

10. $10\frac{1}{2}$ cm

Name _____

Other Metric Units of Length

Name the best unit to measure each distance.
Write *centimeter*, *meter*, or *kilometer*.

1. distance between your eyes _____

2. distance around your head _____

3. height of a building _____

4. distance between two cities _____

5. length of a room _____

Complete each chart.

6.

Meters	1	2		4	5
Centimeters	100		300		

7.

Kilometers	1	2	3		5
Meters	1,000			4,000	

Complete. Use the patterns shown in the charts.

8. 1 m = _____ cm

9. 1 km = _____ m

10. 800 cm = _____ m

11. 9 km = _____ m

12. 10 m = _____ cm

13. 10,000 m = _____ km

© Calvert School

Name _____

Metric Units of Capacity

Complete.

1. 1 L = _____ mL

2. 8,000 mL = _____ L

Does each object hold more or less than 1 liter? Write *more* or *less*.
Use the benchmarks in Section 9.9 of your textbook if needed.

3. soup ladle _____

4. juice glass _____

5. car's gas tank _____

6. medicine dropper _____

7. steam iron _____

8. swimming pool _____

Complete the chart.

9.

Liters	1	2	3	4		6
Milliliters	1,000				5,000	

Solve. Remember to label your answer.

10. Many soft drinks come in 2-liter bottles. How many milliliters are in three 2-liter bottles?

_____

MIXED Practice

Solve. Remember to label your answers.

1. Mrs. Buck has $200. How many ten-dollar bills can she get for her money?

2. Eric caught a fish that weighed 146 ounces and Neil caught a fish that weighed 153 ounces. How much more did Neil's fish weigh?

Name _____

Metric Units of Mass

Complete.

1. 1 kilogram = _____ grams

2. 10,000 g = _____ kg

Is the mass of each object *about* 1 gram or *more than* 1 gram? Write *about* or *more*. Use the benchmarks in Section 9.10 of your textbook if needed.

3. table _____

4. penny _____

5. baseball bat _____

6. piece of paper _____

Name the best unit to measure each mass. Write *gram* or *kilogram*.

7. tennis ball _____

8. desktop computer _____

9. apple _____

10. set of encyclopedias _____

Ring the best measurement.

11. crayon   **a.** 5 g   **b.** 5 kg

12. thick book   **a.** 1 g   **b.** 1 kg

13. large dog   **a.** 35 g   **b.** 35 kg

14. ruler   **a.** 7 g   **b.** 7 kg

Complete the chart.

15.

Kilograms	1	2	3		5
Grams	1,000	2,000		4,000	

Name _____

Tall Tales, Metric Style

In the following problems, you will learn about some legendary, larger-than-life characters from tall tales. A tall tale is a fictional story about a historical figure.

Metric Conversions	
Length	**Capacity**
1 cm = 10 mm	1 liter = 1,000 mL
1 m = 100 cm	**Mass (Weight)**
1 km = 1,000 m	1 kg = 1,000 g

Use the conversion chart to help you answer questions about these fascinating characters.

1. Paul Bunyan was a leader among lumberjacks. It is said that he dug the Grand Canyon. In Minnesota, there is a statue of Paul Bunyan which measures about 550 centimeters tall. Is this statue taller or shorter than a meter?

   _____

2. Paul Bunyan's sidekick was Babe the Blue Ox. There is a 3-meter tall statue of Babe next to Paul's statue in Minnesota. An ox like Babe could eat about 1,400 kilograms of grass each day. How many grams is 1,400 kilograms?

   _____

3. Pecos Bill was a cowboy who was raised by coyotes. Two of his claims to fame were creating the Painted Desert and digging the Rio Grande River. Pecos Bill was so large that he could drink 85 liters of water. About how many milliliters of water could he drink?

_____

4. Johnny Appleseed is well-known for walking across America to plant apple orchards. Today, it takes nearly 40 apples to make 4 liters of apple cider (which is only a little more than 1 gallon). How many milliliters of apple cider is that?

_____

5. John Henry helped build the railroad in the early days of the American West. He could lift 5 tons of steel with one hand—almost 5,000,000 grams! About how much steel could John Henry lift in kilograms?

_____

Name _____

Chapter 9 Review

**Measure the length of each line to the nearest centimeter.
Write the measurement on the line.**

1. _____   2. _____

**Measure the length of each line to the nearest half-inch.
Write the measurement on the line.**

3. _____   4. _____

**Is the mass of each object *about* 1 gram or *more than* 1 gram?
Write *about* or *more*.**

5. calculator _____   6. postage stamp _____

Ring the letter of the best unit to measure the mass of each object.

7.   a. ounce   8. 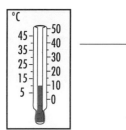   a. ounce
                          b. pound   b. pound

Write the letter of the units used for each measurement.

9. temperature _____   a. liter, milliliter

10. mass _____   b. gram, kilogram

11. length _____   c. degrees Celsius

12. capacity _____   d. centimeter, meter

Write each temperature.

13. 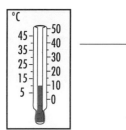 _____   14. 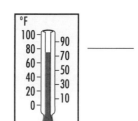 _____   15. 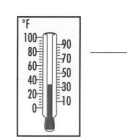 _____

Write the units of measure in order from least to greatest.

16. pound, ton, ounce

    ———————, ———————, ———————

17. meter, kilometer, centimeter

    ———————, ———————, ———————

18. cup, gallon, quart, pint

    ———————, ———————, ———————, ———————

Complete each chart.

19.

Liters	1	5	8	11	13
Milliliters	1,000				

20.

Feet	1	2	3	4	5
Inches	12				

21.

Meters	1,000				
Kilometers	1	3	5	7	9

22.

Quarts	4				
Gallons	1	2	3	6	9

23.

Pounds	1	2	3	4	5
Ounces	16				

Name _____

Chapter 9 Test Prep

Ring the letter of the correct answer.

1. If 4 quarts equal 1 gallon, how many gallons equal 44 quarts?

   a. 8 gal   b. 9 gal   c. 10 gal   d. 11 gal

2. What is the best measurement for the length of a postage stamp?

   a. 1 in.   b. 1 ft   c. 1 yd   d. 1 mi

3. Which of the following is in order from longest to shortest?

   a. meter, centimeter, kilometer   b. centimeter, meter, kilometer

   c. kilometer, meter, centimeter   d. kilometer, centimeter, meter

4. A good benchmark for a yard is a _____.

   a. shoe   b. backyard   c. door width   d. book

5. How many cups are in 1 pint?

   a. 1 cup   b. 2 cups   c. 3 cups   d. 4 cups

6. If 2,000 pounds equal 1 ton, how many pounds equal 5 tons?

   a. 400 pounds

   b. 1,000 pounds

   c. 5,000 pounds

   d. 10,000 pounds

7. What is the best estimate for the mass of a pair of tennis shoes?

   a. 1 gram   b. 10 grams   c. 1 kilogram   d. 10 kilograms

8. What activity would be best if the temperature was 70°F?

   a. ice skating

   b. flying a kite

   c. wearing your swimsuit at the beach

   d. sledding

9. What activity is best if the temperature is 0°C?

   a. ice skating

   b. flying a kite

   c. wearing your swimsuit at the beach

   d. gardening

10. What unit is best for measuring the length of a driveway?

   a. pounds   b. inches   c. yards   d. gallons

Name _____

Points, Lines, and Line Segments

Write the letter of the word(s) that complete each sentence.

_____ 1. A _____ goes on forever
in both directions.

_____ 2. A _____ tells an exact place.

_____ 3. A straight path between two points is
called a _____ .

_____ 4. A line segment has two _____ .

a. arrows
b. endpoints
c. line
d. line segment
e. point

Draw and label each figure.

5. $\overleftrightarrow{TU}$

6. Y
●

7. $\overline{WX}$

Does the drawing show a line segment? Write *yes* or *no*.

8.

9.

10.

11.

_____   _____   _____   _____

Write the number of line segments in each figure.

12.

13.

14.

15.

_____   _____   _____   _____

Name _____

Rays and Angles

Write the letter of the word that completes each sentence.

_____ 1. A(n) _____ begins at one endpoint and goes on forever in one direction.

_____ 2. A(n) _____ is formed when two rays share an endpoint.

_____ 3. A(n) _____ angle is a square angle.

a. acute
b. angle
c. obtuse
d. ray
e. right

Draw and label each figure.

4. right angle ABC

5. $\overrightarrow{DC}$

6. obtuse angle

Use the corner of a piece of paper to name each angle.
Write *acute,* *right,* **or** *obtuse.*

7.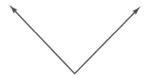

_____

8.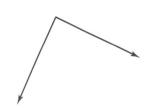

_____

9.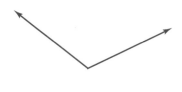

_____

10.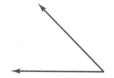

_____

11.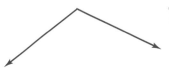

_____

12.

_____

Name _____

Polygons

Decide whether each figure is a polygon. Write *yes* or *no*.

1.

2.

3.

4.

_____   _____   _____   _____

5. Ring the words that describe a polygon. Then explain what makes a figure a polygon.

closed	3-D	plane	crossing lines
curved	open	straight edges	lines do not cross

_____

_____

6. Draw these figures. Use a straightedge to make lines.

3 polygons **(each has a different number of sides)**	**3 figures that are <u>not</u> polygons**

Name _____

Quadrilaterals

**Name each quadrilateral. Write *rectangle, square, trapezoid,* or *rhombus.*
Use each word only once. Then, ring the parallelograms.**

1.   2.   3.   4.

_____   _____   _____   _____

5. Look at the shapes you did not ring. Why are they not
   parallelograms?

   _____

Write the number of right angles in each quadrilateral.

6. square _____   7. rectangle _____

Write the number of line segments in each quadrilateral.

8. trapezoid _____   9. parallelogram _____   10. square _____

11. Pick two of the following shapes and describe how they
    are similar to each other.

    | rhombus | rectangle | square |
    | parallelogram | trapezoid | |

    I chose _____ and _____.

    They are similar because: a. _____.

    b. _____.

Name _____

Triangles

Use the word box to name each triangle. Then write a sentence to describe it.

isosceles	scalene	equilateral
right	acute	obtuse

1. _____
   _____

2.  _____
   _____

3. _____
   _____

Some triangles have right angles. You can test a triangle to see whether it is a right triangle by putting the corner of a piece of paper in the angles of the triangle.

Decide whether each triangle is a right triangle. Write *yes* or *no*.

4.   5.   6.   7.

_____   _____   _____   _____

MIXED Practice

Write the amount of money.

1.

2.

_____   _____

Name _____

Solid Figures

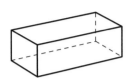   

rectangular   sphere   cube cylinder   pyramid   cone
prism

Name each solid.

1. four triangular surfaces and one square surface

_____

2. one curved surface and no flat surfaces

_____

3. one curved surface and one flat surface

_____

4. six square surfaces

_____

5.

_____

6.

_____

7.

_____

MIXED Practice

Ring the greater amount.

1. 1 c   or   1 pt

2. 1 qt   or   1 pt

3. 6 c   or   1 gal

4. 1 gal   or   3 qt

Name _____

Problem-Solving Strategy: Using Logical Reasoning

Ring the figure in each group that does not belong. Explain your choice.

1. Explain: _____ _____ _____	**2.** Explain: _____ _____ _____
3. Explain: _____ _____ _____	**4.** Explain: _____ _____ _____
5. Explain: _____ _____ _____	**6.** Design your own. Ask a friend to solve.

Name _____

Congruent Figures

On the grid below, draw a figure that is congruent to each given figure. Use a straightedge.

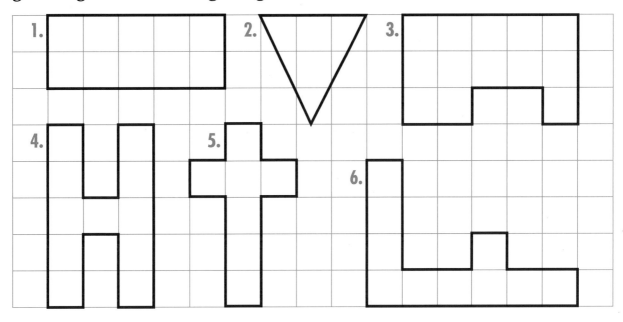

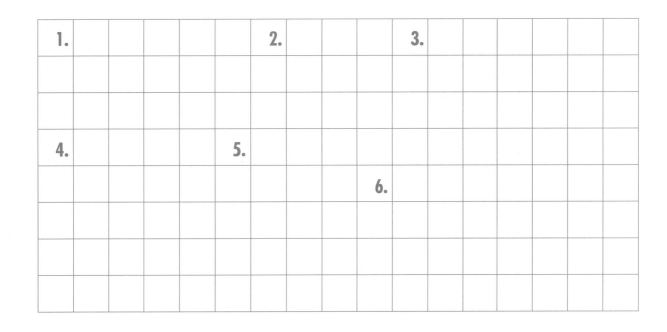

Write the letters of the two congruent figures.

7.

  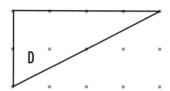

_____ and _____

8.

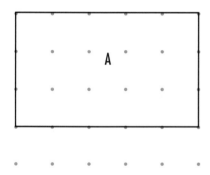 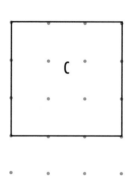

_____ and _____

MIXED Practice

Complete.

1. 2,000 m = _____ km

2. _____ ft = 1 yd

3. _____ T = 4,000 lb

4. 2 kg = _____ g

Name _____

Diagonals

A *diagonal* is a line that joins two corners of a figure.
It must skip at least one corner.

These drawings show diagonals.

 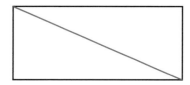

Draw a diagonal in each figure that will divide it into congruent parts.

1.

2.

Draw a diagonal in each figure that will divide it into parts that are not congruent.

3.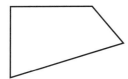

4.

5.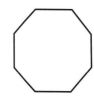

6. Can a triangle have a diagonal? _____ Explain.

_____

_____

7. Can a circle have a diagonal? _____ Explain.

_____

_____

Name _____

Symmetry

Is each dashed line a line of symmetry? Write *yes* or *no*.

1.
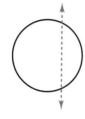
_____

2.
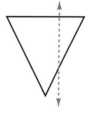
_____

3.
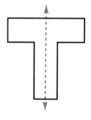
_____

4.

_____

5.
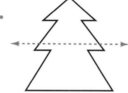
_____

6.
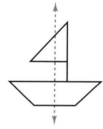
_____

7.

_____

8.
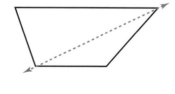
_____

Use a straightedge to draw a line of symmetry for each figure.

9.

10.

11.
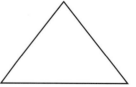

Some figures have more than one line of symmetry. Draw all lines of symmetry for each figure.

12.

13.

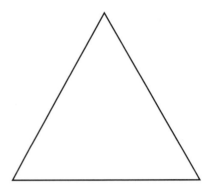

14.

15.

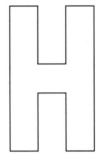

MIXED Practice

Solve. Remember to label your answers.

1. Erica has a rectangular poster that is 12 inches wide and 16 inches long. Draw a model of the poster and label the sides.

2. A book store sold 79 mystery books on Friday, 87 on Saturday, and 54 on Sunday. How many mystery books were sold during the three days?

Name _____

What is my Name?

Each dotted line shows a line of symmetry in a letter. Draw the other half of the letter to write a person's name.

1.

2.

3.

4.

5.

6. Write other capital letters that are symmetrical.

_____

Name _____

Slide, Flip, and Turn

How was each figure moved? Write *slide*, *flip*, or *turn*.

1. 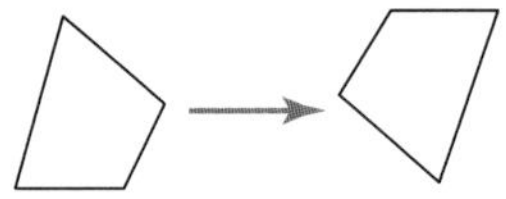 ▷ → ◺ _____

2. 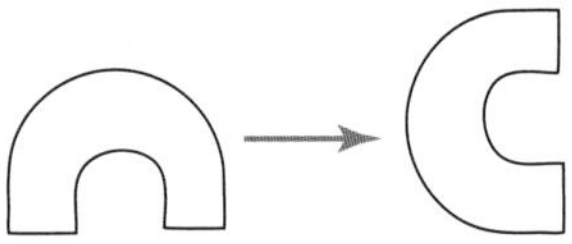 ∩ → C _____

3. 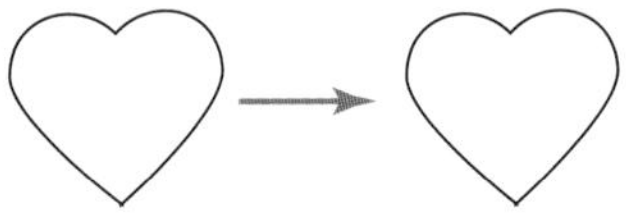 ♡ → ♡ _____

4. 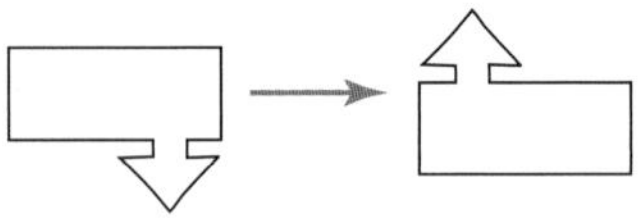 ⬐ → ⬆ _____

5. 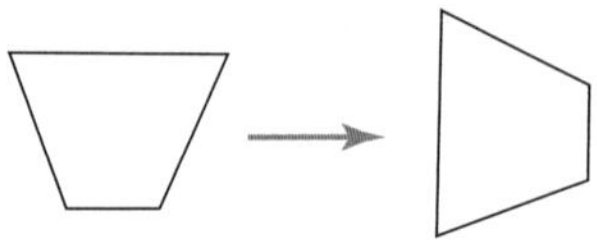 ▽ → ▷ _____

6. 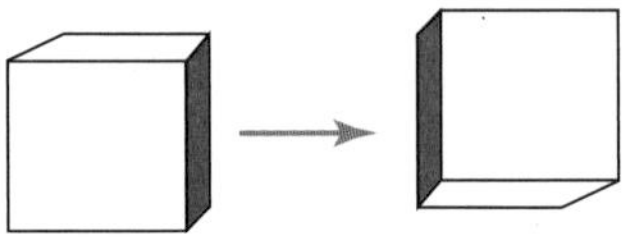 ▭ → ◻ _____

**Use this shape as your original each time.
Draw each of the following.**

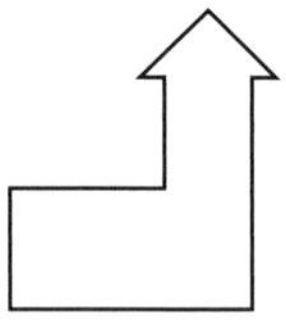

⬆

7. 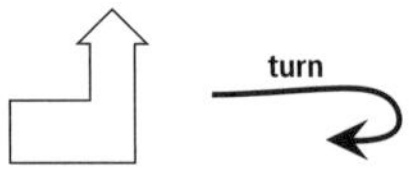 ⬏ turn ↶ _____

8. 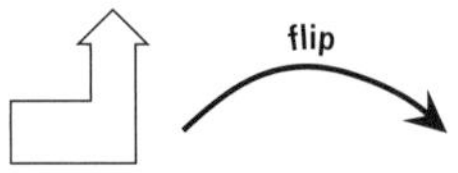 ⬏ flip → _____

9. 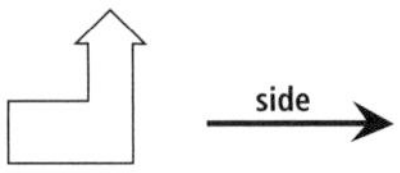 ⬏ side → _____

Name _____

Perimeter

Find the perimeter of each shape.

1.

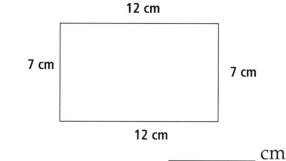

12 cm

7 cm   7 cm

12 cm

_____ cm

2.

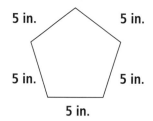

5 in.   5 in.

5 in.   5 in.

5 in.

_____ in.

3.

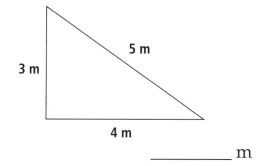

5 m

3 m

4 m

_____ m

4.

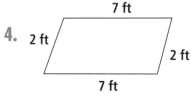

7 ft

2 ft   2 ft

7 ft

_____ ft

Complete the chart.

5. A movie poster is 18 inches long and 12 inches wide.
   A music poster is 15 inches long and 14 inches wide.
   Show each poster's measurements and perimeter in the chart.

Movie Posters

Poster	Length (in.)	Width (in.)	Perimeter (in.)
Movie			
Music			

Which poster has the greater perimeter? _____

Name _____

Area

Write the area of each figure in square centimeters.

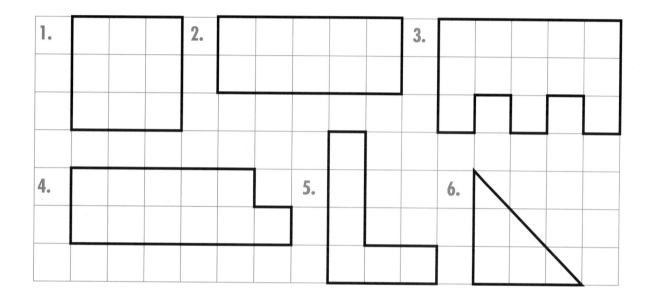

1. _____   2. _____   3. _____

4. _____   5. _____   6. _____

7. Draw four figures that each have an area of 8 square centimeters.

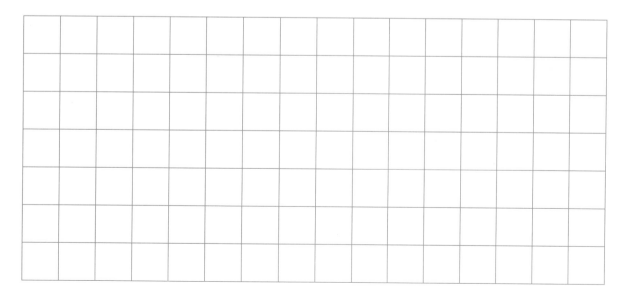

8. Draw five figures that each have an area of 12 square centimeters.

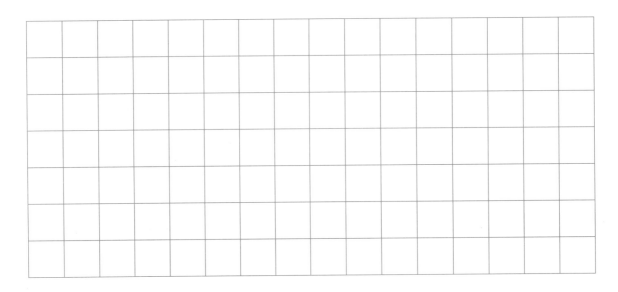

MIXED Practice

Write how many hundreds, tens, and ones.

1. 683 = _____ hundreds _____ tens _____ ones

2. 407 = _____ tens _____ hundreds _____ ones

3. 271 = _____ ones _____ tens _____ hundreds

Estimate by rounding.

4.   476
  + 315

5.   83,491
  − 46,801

6.   15,459
  + 34,819

Name _____

Area

Draw as many shapes as possible with an area of 18 square centimeters. Next to each write the perimeter (the distance around the outside of the figure).

Name _____

Volume

Use cubes to make each figure. Write the volume of each figure.

1.

_____ cubic units

2.

_____ cubic units

3.

_____ cubic units

4.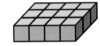

_____ cubic units

5.

_____ cubic units

6.

_____ cubic units

Write the volume of each figure in cubic centimeters.

7.

_____ cubic cm

8.

_____ cubic cm

9.

_____ cubic cm

MIXED Practice

Solve. Remember to label your answers.

1. Use the numbers to write the fact family: 9, 16, 7.

_____

_____

_____

2. Joe left the store at 6:07 P.M. He arrived home at 6:40 P.M. About how long did it take him to reach home?

_____

Name _____

Problem-Solving Application: Tiling Patterns

Tiling patterns have no holes or gaps.

Use these pattern blocks to make the tiling patterns described.

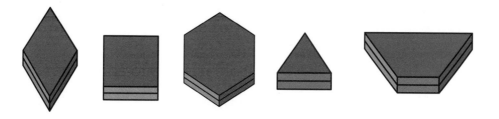

Make a tiling pattern with the following:

1. 2 different kinds of pattern blocks

2. 3 different kinds of pattern blocks

3. all 5 kinds of pattern blocks

Name _____

Chapter 10 Review

Write the letter of the best answer.

_____ **1.** an exact place

_____ **2.** a straight path between 2 points

_____ **3.** figures with the same size and shape

_____ **4.** like a square corner

_____ **5.** sphere

> **a.** congruent
> **b.** line segment
> **c.** point
> **d.** right angle
> **e.** solid figure

Identify the solid.

6. six flat surfaces that are all squares

_____

7. one curved surface and two flat surfaces

_____

Does the drawing show a line segment? Write *yes* or *no*.

8.

9.

10.

11.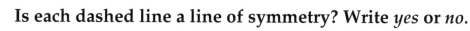

_____   _____   _____   _____

Is each dashed line a line of symmetry? Write *yes* or *no*.

12.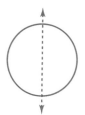

13.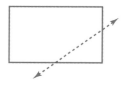

14.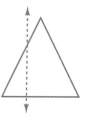

_____   _____   _____

Name each polygon.

15.

16.

17.

18.

_____   _____   _____   _____

Name each solid figure.

19.

20.

21.

22.

_____   _____   _____   _____

23. Ring the figure that is congruent to the first one.

24. Find the perimeter of the figure in centimeters. Then find the area of the figure in square centimeters.

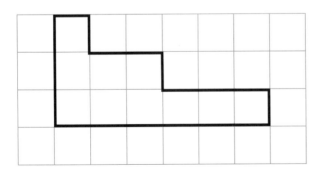

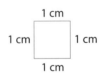

P = _____ cm

A = _____ sq cm

© Calvert School

Name _____

Chapter 10 Test Prep

Ring the letter of the correct answer.

1. These shapes are _____.

   a. quadrilaterals   **b.** triangles

   c. open   **d.** congruent

2. Two rays that share an endpoint make a(n) _____.

   a. line   **b.** line segment   **c.** double ray   **d.** angle

3. This movement shows _____.

   a. a flip   **b.** a slide

   c. a turn   **d.** none of these

4. This triangle is _____.

   a. equilateral   **b.** right   **c.** scalene   **d.** isosceles

5. These objects are all _____.

   a. polygons   **b.** solid figures   **c.** congruent   **d.** symmetrical

6. How many lines of symmetry does this shape have?

   a. 0   **b.** 1

   c. 2   **d.** 3

7. These two figures are _____.

   a. congruent   **b.** symmetrical

   c. turned   **d.** equilateral

8. What is the volume?

   a. 8 cubic units   **b.** 10 cubic units

   c. 14 cubic units   **d.** 18 cubic units

9. What is the area?

   a. 19 square units **b.** 20 square units

   c. 21 square units **d.** 22 square units

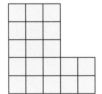

10. What is the perimeter?

    a. 18 cm   **b.** 20 cm

    c. 25 cm   **d.** 30 cm

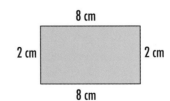

Name _____

Fractions as Part of a Whole

Use the words *numerator* or *denominator* to complete each sentence.

1. The top number in a fraction is called the _____.
   It tells how many equal parts are being considered.

2. The bottom number in a fraction is called the _____.
   It tells how many equal parts are in the whole.

Write the fraction for the shaded parts.

3.

4.

5.

6.

Write each fraction.

7. three fifths _____   8. four eighths _____   9. one half _____

Write the letter of the drawing for each fraction. Color the parts named.

a.   b.   c.

10. $\frac{2}{3}$ _____   11. $\frac{3}{8}$ _____   12. $\frac{1}{4}$ _____

Solve.

13. The Jones family ate $\frac{7}{8}$ of a pizza. Write the fraction that tells what part was left.

14. A graham cracker is split into 4 equal parts. Susan ate 1 of the parts. What fraction of the cracker did she eat?

_____

_____

Name _____

Fractions as Part of a Set

Complete.

1. How many circles are yellow? _____

2. How many circles are there in all? _____

3. Write the fraction that shows the part of the group that is yellow. _____

Write each fraction.

4.

   glasses with orange juice

5.

   shirts with stripes

$$\dfrac{\boxed{}}{\boxed{}}$$

$$\dfrac{\boxed{}}{\boxed{}}$$

Complete.

6. Color $\frac{2}{5}$ of the dogs brown.

7. Color $\frac{1}{6}$ of the balloons green, $\frac{1}{6}$ red, $\frac{2}{6}$ blue, $\frac{1}{6}$ yellow, and $\frac{1}{6}$ orange.

MIXED Practice

Solve. Remember to label your answers.

1. What is the value of 9 nickels?

2. One book costs $5. How much do ten books cost?

_____   _____

Name _____

Fractions

1. Choose and draw a pattern block. If this block shows $\frac{1}{4}$, draw a picture that shows 1 whole.

2. Choose and draw a different pattern block. If this block shows $\frac{1}{3}$, draw a picture that shows 2 wholes.

Name _____

Comparing and Ordering Fractions

Compare the shaded parts. Write <, >, or =.

1.

$\frac{1}{4}\ \bigcirc\ \frac{2}{4}$

2.

$\frac{4}{6}\ \bigcirc\ \frac{3}{6}$

3.

$\frac{2}{8}\ \bigcirc\ \frac{1}{4}$

4.

$\frac{5}{10}\ \bigcirc\ \frac{2}{5}$

5.

$\frac{2}{4}\ \bigcirc\ \frac{5}{8}$

6.

$\frac{1}{2}\ \bigcirc\ \frac{1}{3}$

Color the parts named to prove each statement.

7. $\frac{1}{2} = \frac{2}{4}$

8. $\frac{2}{6} < \frac{1}{2}$

Solve.

9. Jim and Elaine are reading the same book. Jim is $\frac{5}{8}$ of the way through the book and Elaine is $\frac{1}{2}$ of the way through. Who has read more? (Hint: Make a drawing.)

_____

10. Minnie made these drawings to compare $\frac{2}{2}$ and $\frac{1}{2}$. Are her drawings correct? Explain.

_____

_____

Name _____

Comparing Fractions

Use the fraction chart to compare. Write <, >, or =.

1. $\dfrac{1}{6}$ ◯ $\dfrac{3}{8}$ 2. $\dfrac{5}{6}$ ◯ $\dfrac{1}{4}$ 3. $\dfrac{3}{4}$ ◯ $\dfrac{6}{8}$ 4. $\dfrac{2}{6}$ ◯ $\dfrac{3}{8}$

5. $\dfrac{2}{8}$ ◯ $\dfrac{1}{4}$ 6. $\dfrac{1}{2}$ ◯ $\dfrac{5}{6}$ 7. $\dfrac{1}{2}$ ◯ $\dfrac{5}{10}$ 8. $\dfrac{2}{3}$ ◯ $\dfrac{1}{2}$

9. $\dfrac{4}{5}$ ◯ $\dfrac{5}{10}$ 10. $\dfrac{2}{10}$ ◯ $\dfrac{1}{3}$ 11. $\dfrac{3}{5}$ ◯ $\dfrac{1}{2}$ 12. $\dfrac{8}{10}$ ◯ $\dfrac{4}{5}$

Fraction Chart

$\frac{1}{2}$					$\frac{1}{2}$				
$\frac{1}{3}$			$\frac{1}{3}$			$\frac{1}{3}$			
$\frac{1}{4}$		$\frac{1}{4}$		$\frac{1}{4}$			$\frac{1}{4}$		
$\frac{1}{5}$		$\frac{1}{5}$		$\frac{1}{5}$		$\frac{1}{5}$		$\frac{1}{5}$	
$\frac{1}{6}$		$\frac{1}{6}$		$\frac{1}{6}$		$\frac{1}{6}$		$\frac{1}{6}$	$\frac{1}{6}$
$\frac{1}{8}$	$\frac{1}{8}$	$\frac{1}{8}$	$\frac{1}{8}$	$\frac{1}{8}$	$\frac{1}{8}$	$\frac{1}{8}$	$\frac{1}{8}$		
$\frac{1}{10}$	$\frac{1}{10}$	$\frac{1}{10}$	$\frac{1}{10}$	$\frac{1}{10}$	$\frac{1}{10}$	$\frac{1}{10}$	$\frac{1}{10}$	$\frac{1}{10}$	$\frac{1}{10}$

Name _____

Equivalent Fractions

Equivalent fractions show <u>equal</u> amounts.

Complete.

1.

$$\frac{1}{2} = \frac{}{6}$$

2.

$$\frac{3}{5} = \frac{}{10}$$

3.

$$\frac{6}{8} = \frac{}{4}$$

4.

$$\frac{1}{3} = \frac{}{6}$$

5.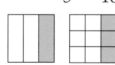

$$\frac{1}{3} = \frac{3}{}$$

6.

$$\frac{2}{5} = \frac{4}{}$$

7.

$$\frac{1}{2} = \frac{}{4}$$

8.

$$\frac{2}{4} = \frac{}{8}$$

9.

$$\frac{4}{8} = \frac{}{16}$$

10. Look for a pattern in problems 7–9. Then complete the following problems.

$$\frac{1}{2} = \frac{}{6} \qquad \frac{1}{2} = \frac{}{10} \qquad \frac{1}{2} = \frac{}{20}$$

Complete.

11. $\frac{}{2} = 1$

12. $\frac{3}{} = 1$

13. $\frac{}{6} = 1$

© Calvert School

Solve. Remember to label your answers.

14. Buddy told his friends he had eaten $\frac{2}{2}$ of a candy bar. How much did he eat?

_____

15. June ran $\frac{1}{2}$ of a mile. Judy ran $\frac{2}{4}$ of a mile. Judy said she ran farther. Was she right? Explain.

_____

_____

MIXED Practice

Multiply or divide.

1. $9\overline{)83}$ **2.** $7\overline{)52}$ **3.** $5\overline{)39}$

4. $(3 \times 3) \times 6 =$ _____   **5.** $56 = \boxed{} \times 7$

Write the coordinates for each point.

6. E _____   B _____

Name the points with the given coordinates.

7. (1, 5) _____   (5, 0) _____

Plot and label each point on the grid.

8. F (3, 4)   G (0, 3)

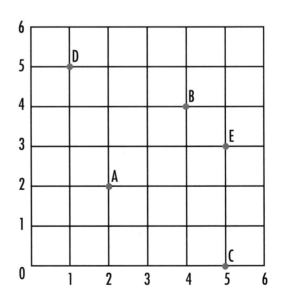

Name _____

More Equivalent Fractions

Use a straightedge and the fraction chart to find
an equivalent fraction for each.

Fraction Chart

1 Whole											
$\frac{1}{2}$						$\frac{1}{2}$					
$\frac{1}{3}$				$\frac{1}{3}$				$\frac{1}{3}$			
$\frac{1}{4}$			$\frac{1}{4}$			$\frac{1}{4}$			$\frac{1}{4}$		
$\frac{1}{5}$		$\frac{1}{5}$		$\frac{1}{5}$		$\frac{1}{5}$		$\frac{1}{5}$			
$\frac{1}{6}$		$\frac{1}{6}$		$\frac{1}{6}$		$\frac{1}{6}$		$\frac{1}{6}$		$\frac{1}{6}$	
$\frac{1}{7}$	$\frac{1}{7}$		$\frac{1}{7}$		$\frac{1}{7}$		$\frac{1}{7}$		$\frac{1}{7}$		$\frac{1}{7}$
$\frac{1}{8}$	$\frac{1}{8}$	$\frac{1}{8}$	$\frac{1}{8}$	$\frac{1}{8}$		$\frac{1}{8}$		$\frac{1}{8}$		$\frac{1}{8}$	
$\frac{1}{9}$	$\frac{1}{9}$	$\frac{1}{9}$	$\frac{1}{9}$	$\frac{1}{9}$	$\frac{1}{9}$	$\frac{1}{9}$		$\frac{1}{9}$		$\frac{1}{9}$	
$\frac{1}{10}$	$\frac{1}{10}$	$\frac{1}{10}$	$\frac{1}{10}$	$\frac{1}{10}$	$\frac{1}{10}$	$\frac{1}{10}$	$\frac{1}{10}$	$\frac{1}{10}$	$\frac{1}{10}$		
$\frac{1}{11}$	$\frac{1}{11}$	$\frac{1}{11}$	$\frac{1}{11}$	$\frac{1}{11}$	$\frac{1}{11}$	$\frac{1}{11}$	$\frac{1}{11}$	$\frac{1}{11}$	$\frac{1}{11}$	$\frac{1}{11}$	
$\frac{1}{12}$	$\frac{1}{12}$	$\frac{1}{12}$	$\frac{1}{12}$	$\frac{1}{12}$	$\frac{1}{12}$	$\frac{1}{12}$	$\frac{1}{12}$	$\frac{1}{12}$	$\frac{1}{12}$	$\frac{1}{12}$	$\frac{1}{12}$

1. $\frac{4}{10}$ _____

2. $\frac{10}{10}$ _____

3. $\frac{2}{3}$ _____

4. $\frac{4}{5}$ _____

5. $\frac{6}{8}$ _____

6. $\frac{2}{8}$ _____

7. $\frac{2}{6}$ _____

8. $\frac{4}{12}$ _____

9. $\frac{3}{9}$ _____

Use the fraction chart. Write the letter that shows the equivalent fraction.

___ 10. $\frac{6}{12}$ a. $\frac{4}{5}$

___ 11. $\frac{8}{10}$ b. $\frac{2}{10}$

___ 12. $\frac{1}{4}$ c. $\frac{2}{3}$

___ 13. $\frac{7}{7}$ d. $\frac{1}{2}$

___ 14. $\frac{1}{5}$ e. $\frac{3}{4}$

___ 15. $\frac{5}{6}$ f. $\frac{10}{12}$

___ 16. $\frac{6}{9}$ g. $\frac{11}{11}$

___ 17. $\frac{6}{8}$ h. $\frac{3}{12}$

Complete.

18.

$$\frac{1}{2} = \frac{}{6}$$

19.

$$\frac{1}{3} = \frac{}{6}$$

20.

$$\frac{1}{2} = \frac{}{10}$$

21.

$$\frac{2}{3} = \frac{}{6}$$

Name _____

Problem-Solving Application: Using Any Strategy

Use any strategy to solve. Remember to label your answers.

1. Continue the pattern.

2. Joe had some baseball cards. After Sue gave him 25 more, he had 137 cards in all. How many cards did Joe have in the beginning?

   _____

3. Allen used 3 tablespoons of sugar for each batch of muffins. He made 4 batches. How much sugar did he use in all for 4 batches?

   _____

4. Two children can go to the movies for $7. How much will it cost for 6 children to go to the movies?

   _____

5. On Saturday, Maria sold 398 cookies. On Sunday, she sold 305 cookies. About how many cookies did she sell that weekend?

   _____

6. Evan has blue pants and black pants. He has a white shirt, a yellow shirt, and a tan shirt. How many different outfits can he make?

   _____

Name _____

Finding Fractional Parts of a Set

$\frac{2}{4}$ of 12  $\frac{2}{3}$ of 9

$\frac{1}{4}$ of 12 = 3   1 of the 4 groups = 3 $\frac{1}{3}$ of 9 = 3, so $\frac{2}{3}$ of 9 = 6

$\frac{2}{4}$ of 12 = 6   2 of the 4 groups = 6

Complete.

1. Finding $\frac{1}{5}$ of 25 is like dividing 25 by _____.

2. Finding $\frac{1}{2}$ of a number is like dividing that number by _____.

3.

   $\frac{2}{5}$ of 10 = _____

4.

   $\frac{1}{3}$ of 12 = _____

5.

   $\frac{2}{3}$ of 15 = _____

6.

   $\frac{1}{2}$ of 10 = _____

7.

   $\frac{3}{4}$ of 16 = _____

8.

   $\frac{2}{2}$ of 18 = _____

Complete. Use counters to check your answer.

9. $\frac{1}{6}$ of 6 = _____

10. $\frac{1}{4}$ of \$20 = _____

11. $\frac{2}{5}$ of \$25 = _____

12. $\frac{2}{2}$ of \$14 = _____

13. $\frac{1}{3}$ of 24 = _____

14. $\frac{3}{7}$ of 35 = _____

15. $\frac{3}{4}$ of 8 = _____

16. $\frac{2}{3}$ of 12 = _____

Solve. Remember to label your answers.

17. When Mrs. Myers ordered tickets to a show, she had to send half the total cost right away. The tickets cost \$18. How much did Mrs. Myers send?

18. Barbara bought a dozen eggs. When she got home from the store she found that $\frac{1}{3}$ of the eggs were cracked. How many were cracked?

19. Two thirds of a dozen donuts have icing. How many donuts have icing?

20. Three fourths of a foot equals how many inches?

MIXED Practice

Write the coordinates for each point.

1. A _____

2. B _____

3. C _____

Name _____

Improper Fractions and Mixed Numbers

Ring the letter of the correct answer.

1. Another way to write $4\frac{2}{3}$ is _____.

   a. $4 - \frac{2}{3}$ b. $\frac{42}{3}$

   c. $4 + \frac{2}{3}$ d. $\frac{2}{3}$

2. The picture shows _____.

   a. $\frac{3}{2}$ b. $3\frac{1}{2}$

   c. $3\frac{3}{2}$ d. $\frac{2}{3}$

Write the mixed number.

3. two and three fourths _____

4. six and seven tenths _____

5. one and one half _____

6. three and two fifths _____

Write an improper fraction and a mixed number to name the shaded part.

7.

___ ___

8.

___ ___

9.

___ ___

10.

___ ___

Shade each picture to show the amount.

11. $3\frac{1}{8}$

12. $\frac{9}{4}$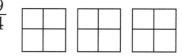

Solve. Remember to label your answers.

13. Ian bought 2 dozen donuts and Peg bought $1\frac{1}{2}$ dozen. Who bought more donuts?

14. How many minutes are in $1\frac{1}{2}$ hours?

_____   _____

Name _____

Adding and Subtracting Fractions

Add or subtract.

1. $\frac{2}{5} + \frac{1}{5} =$ ___

2. $\frac{3}{4} + \frac{1}{4} =$ ___

3. $\frac{9}{10} - \frac{9}{10} =$ ___

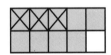

4. $\frac{5}{6} - \frac{2}{6} =$ ___

5. $\frac{8}{10} - \frac{4}{10} =$ ___

6. ___ $= \frac{1}{3} + \frac{1}{3}$

7. ___ $= \frac{6}{7} - \frac{1}{7}$

8. $\frac{7}{9} - \frac{5}{9} =$ ___

9. $\frac{12}{12} - \frac{5}{12} =$ ___

10. $\frac{5}{15} + \frac{7}{15} =$ ___

Solve. Remember to label your answers.

11. At a party on Saturday, a family ate $\frac{3}{8}$ of a cake. On Sunday night they ate $\frac{2}{8}$ more of the cake. How much cake did they eat in all?

_____

How much cake was left?

_____

12. Derek ran $\frac{7}{8}$ of a mile on Monday. He only ran $\frac{4}{8}$ of a mile on Wednesday. How much more did he run on Monday than on Wednesday?

_____

Name _____

Problem-Solving Skill: Identifying Extra Information

Underline the extra information. Then solve.

1. Mrs. Vega bought 2 dozen eggs for $1.90 and 4 dozen donuts for $7.00. How much did she spend in all?

   _____

2. Mr. Neal sold 144 car tires for $7,200 and 219 truck tires for $15,330. How many more truck tires did he sell than car tires?

   _____

3. Barbara spent $\frac{1}{5}$ of her summer at camp, $\frac{2}{5}$ of her summer at her grandparents' house, $\frac{1}{5}$ of her summer at the beach, and $\frac{1}{5}$ of her summer at her house. What fraction of the summer did Barbara spend away from home?

   _____

4. Mary has 8 shirts with pink stripes, 2 shirts with blue stripes, and 3 shirts with no stripes. Mary has 4 pairs of shorts. What fraction of the shirts have stripes?

   _____

5. A baseball team is scheduled to play $\frac{5}{22}$ of their games on Thursdays, $\frac{6}{22}$ on Fridays, $\frac{6}{22}$ on Saturdays, and $\frac{5}{22}$ on Sundays. What fraction of the games are scheduled on weekends?

   _____

6. Emily is 9 years old and weighs 72 pounds. Her 12-year-old brother weighs 105 pounds. How much heavier is her brother?

   _____

Name _____

Chapter 11 Review

Write the letter of the correct answer.

1. a whole number and a fraction _____

2. fractions that are equal _____

3. the part of a fraction that tells how many parts are being considered (top number) _____

4. the part of a fraction that tells how many parts are in the whole (bottom number) _____

a. equivalent

b. denominator

c. mixed number

d. numerator

Write a fraction for the shaded parts.

5. _____

6. _____

7.  _____

Write an improper fraction and a mixed number to name the shaded parts.

8.  _____ _____

9.

_____ _____

Complete.

10. $\frac{1}{3}$ of 12 = _____

OOOO
OOOO
OOOO

11. $\frac{3}{8}$ of 16 = _____

□□□□□□□
□□□□□□□

12. _____ = $\frac{1}{2}$ of 10

13. _____ = $\frac{1}{3}$ of 9

Compare the shaded parts. Write <, >, or =.

14.

$$\frac{3}{8} \bigcirc \frac{2}{4}$$

15.

$$\frac{1}{2} \bigcirc \frac{4}{6}$$

Complete.

16.

$$\frac{1}{5} = \frac{}{10}$$

17.

$$\frac{1}{3} = \frac{}{9}$$

18.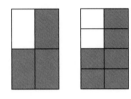

$$\frac{3}{4} = \frac{}{8}$$

Add or subtract.

19. $\frac{5}{7} + \frac{1}{7} =$ ___

20. $\frac{19}{20} - \frac{11}{20} =$ ___

21. $\frac{2}{8} + \frac{1}{8} + \frac{3}{8} =$ ___

22. $\frac{4}{9} + \frac{2}{9} =$ ___

Order from least to greatest.

23. $\frac{6}{8}, \frac{2}{8}, 1\frac{5}{8}$

24. $3\frac{6}{10}, \frac{8}{10}, 2\frac{4}{10}$

___, ___, ___

___, ___, ___

Solve.

25. A group of six children went bowling. Two sixths of the children were girls. Write the fraction that tells what part were boys.

_____

Name _____

Chapter 11 Test Prep

Ring the letter of the correct answer.

1. Order from least to greatest: $\frac{5}{8}, 1\frac{6}{8}, \frac{2}{8}$

   a. $\frac{5}{8}, \frac{2}{8}, 1\frac{6}{8}$ b. $\frac{2}{8}, \frac{5}{8}, 1\frac{6}{8}$

   c. $1\frac{6}{8}, \frac{5}{8}, \frac{2}{8}$ d. none of these

2. Add. $\frac{2}{7} + \frac{3}{7}$

   a. $\frac{1}{7}$ b. $\frac{5}{7}$

   c. $\frac{23}{7}$ d. none of these

3. Subtract. $\frac{8}{9} - \frac{4}{9}$

   a. $\frac{84}{9}$ b. $\frac{12}{9}$

   c. $\frac{4}{9}$ d. none of these

4. Which fraction is equivalent to $\frac{1}{4}$?

   a. $\frac{1}{2}$ b. $\frac{1}{8}$

   c. $\frac{4}{8}$ d. none of these

5. A mixed number can be written as _____.

   a. an improper fraction   b. a numerator

   c. a denominator   d. none of these

6. $\frac{1}{8}$ of 24 is _____ .

   a. 1   b. 3

   c. 5   d. none of these

7. Compare. $\frac{1}{3} \bigcirc \frac{1}{2}$

   a. <   b. >

   c. =   d. none of these

8. There are 6 hamburgers in a package. Darren cooked $\frac{2}{3}$ of the package. How many hamburgers did he cook?

   a. 2 hamburgers   b. 4 hamburgers

   c. 6 hamburgers   d. none of these

9. Sally brought a dozen balloons to a party. One fourth of the balloons were red. How many balloons were red?

   a. 1 balloon   b. 3 balloons

   c. 4 balloons   d. none of these

10. Which shows $\frac{12}{7}$ as a mixed number? (Shade the rectangles for help.)

    a. $\frac{15}{7}$ b. $1\frac{12}{7}$

    c. $1\frac{7}{12}$ d. none of these

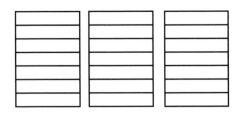

© Calvert School

Name _____

Tenths

Write a fraction and a decimal for each.

1.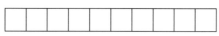
2.
3.

$$\frac{}{10} = \underline{}.\underline{}$$ $$\underline{} = \underline{}.\underline{}$$ $$\underline{} = \underline{}.\underline{}$$

4. seven tenths _____ _____

5. four tenths _____ _____

Shade to show each amount.

6. 0.8

7. 0.2
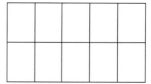

8. 0.4
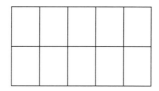

9. 1.0
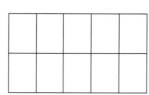

Solve.

10. There were 10 words on a spelling test. Katie spelled 0.9 of the words correctly. Mike spelled 0.8 correctly. Who spelled more words correctly on the spelling test?

_____

11. Dan walked 0.6 mile. Duke walked 0.5 mile. Who walked farther? Explain.

_____

_____

MIXED Practice

Estimate by rounding.

1.   37
   + 49

2.   65
   − 29

3.   193
   − 158

4.   2,868
   + 6,826

Name _____

Hundredths

Complete the chart. Follow the sample.

	Cents	Fraction of a dollar	Money	Decimal
Sample	43 cents	$\frac{43}{100}$	$0.43	0.43
1.	25 cents			
2.	78 cents			
3.	38 cents			
4.	65 cents			

Write the decimal. Think about writing part of a dollar.

5. $\frac{44}{100}$ = _____

6. $\frac{9}{100}$ = _____

7. $\frac{56}{100}$ = _____

8. $\frac{13}{100}$ = _____

9. $\frac{84}{100}$ = _____

10. $\frac{8}{100}$ = _____

11. sixty hundredths = _____

12. five hundredths = _____

Solve. Remember to label your answers.

13. Downhill skiers are timed in hundredths of a second. Ross skied part of a course in $\frac{58}{100}$ of a second. Write the time as a decimal.

_____

14. There are 100 craft sticks in a package. Marnie used 87 sticks for her project. Write as a decimal the part of the package that she used.

_____

Name _____

Comparing Fractions and Decimals

Complete the chart.

	Words	Fraction	Decimal	Draw a picture
1.	two tenths			
2.	seven tenths			
3.	fifty-four hundredths			
4.	nine hundredths			
5.	ninety hundredths			

6. Explain the difference between four tenths and four hundredths.

_____

_____

Name _____

Decimals Greater than One

Write a mixed number and a decimal for each.

1. five and one tenth

2. three and four hundredths

_____ _____

_____ _____

Write a mixed number and a decimal to name the shaded parts.

3.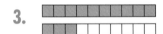

4.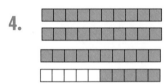

_____ _____

_____ _____

5.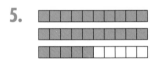

6.

7.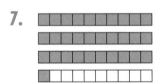

_____ _____

_____ _____

_____ _____

Color the picture to show each number.

8. 2.3

9. 1.8

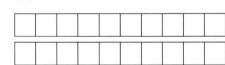

Write the missing numbers to complete each pattern.

10. 1.8, _____, 2.0, 2.1, _____, _____, 2.4

11. 3.32, _____, 3.52, 3.62, _____, 3.82, _____

Name _____

Problem-Solving Skill: Using Logical Reasoning

🎁	=3
📫	=6
❄	=9
✏	=4

Use the chart to solve each problem.

1. (✏ × 🎁) ÷ 📫 = _____

2. (❄ × ✏) ÷ 📫 = _____

3. 🎁 + 📫 + ❄ + ✏ = _____

4. If (🎁 × ❄) − 6 = ⟨×⟩ , what does ⟨×⟩ equal? _____

Solve. Remember to label your answers.

5. Arrange the digits 1, 2, 3, 4, 5, 6, 7, and 8 in two groups so that each group has the same sum.

_____

6. Mrs. Hicks is between 30 and 40 years old. You say her age when you count by 3s, 4s, and 6s. How old is Mrs. Hicks?

_____

7. At the Yogurt Shop, Manny received $3.70 in change from a $5 bill. Use the chart to determine what he bought.

Yogurt Prices	
Small	80¢
Large	$1.49
Toppings	50¢

_____

8. Find a way to make $1.00 using 9 coins.

_____

9. Ring the group of coins that has a sum of 25¢.

   1 dime, 2 nickels, 5 pennies

   2 dimes, 5 nickels

   1 dime, 5 nickels, 5 pennies

10. Write the same digit in each green box to complete the addition problem.

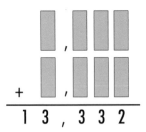

MIXED Practice

Order the numbers from least to greatest.

1. 2,704; 2,740; 2,470

   _____; _____; _____

2. $\frac{5}{8}$, $1\frac{6}{8}$, $\frac{2}{8}$

   _____, _____, _____

Solve. Remember to label your answers.

3. Bob left home at 3:15 to walk to the store. It takes 15 minutes to walk to the store. At what time will Bob arrive at the store?

   _____

4. Walt spent $10.98 for a CD and $4.98 for a book. How much change will he receive from $20?

   _____

Name _____

Comparing and Ordering Decimals

Compare. Write <, >, or =.

1. 3.81 ◯ 3.18
2. 7.3 ◯ 7.03
3. 8.51 ◯ 8.74

4. 3.29 ◯ 3.2
5. 1.79 ◯ 17.9
6. 3.21 ◯ 3.58

Order the decimals from least to greatest.

7. 0.61, 0.71, 0.8, 0.36   _____, _____, _____, _____

8. 24.6, 22.76, 23.15, 24.09   _____, _____, _____, _____

9. 15.93, 15.5, 13.94, 15.25   _____, _____, _____, _____

10. 91.04, 91.84, 91.78, 91.0   _____, _____, _____, _____

Use the chart to complete. Whose fish weighed more? Circle your answer.

Fish-Catching Contest Leaders	
Billy	7.3 kg
Tara	7.12 kg
Cindy	6.99 kg
Veronica	7.41 kg
Justin	7.49 kg
Tracy	7.05 kg
Derek	6.71 kg

11. Billy or Tara
12. Veronica or Justin

13. Tara or Tracy
14. Derek or Cindy

15. Who caught the heaviest fish? _____

16. Order the children from heaviest fish to lightest fish caught.

_____

_____

_____

17. What fraction of the fish caught are over 7 kilograms? _____

Name _____

Order the decimals from least to greatest in the green boxes. Write each decimal's corresponding letter in the white boxes.

Example:

2.1	0.2	2.5	5.71
O	W	R	D

0.2	2.1	2.5	2.75
W	O	R	D

1.

0.23	0.32	0.6	0.57	0.71	0.2	0.51	0.4
E	C	L	A	S	D	M	I

2.

1.21	1.12	1.13
D	A	N

3.

2.92	2.82	2.9	2.85	2.94	3.0	2.8	2.87	2.97
I	R	T	A	O	S	F	C	N

4.

5.17	5.77	5.1	5.71
A	E	H	V

5.

9.22	9.43	9.12	9.21	9.56	9.1	9.7	9.65
E	N	R	I	D	F	Y	L

6.

58.67	67.58	68.57	58.76	65.78	68.75	67.85	56.87	65.87
E	I	N	L	A	S	O	R	T

Write each word from problems #1–6 to reveal how fractions and decimals get along.

_____   _____   _____
        # 1   # 2   # 3

_____   _____   _____
        # 4   # 5   # 6

Name _____

Adding Decimals

Add. Line up the decimal points.

1.   1.23
    + 0.7

2.   34.31
    + 27.75

3. $68.07
    + 14.49

4. 15.9 + 18.9

5. 10.62 + 27.48

6. $37.01 + $3.69

7. 23.84 + 45.7

8. 14.82 + 92

9. $13.35 + $4.92

Check to see if each problem is correct. If the sum is correct, write *yes* in the box. If it is incorrect, rework the problem to find the sum.

10.   14.8
    + 1.28
   —————
    2.76

11.   17.58
    + 24.93
   —————
    42.51

12.   34.08
    + 9.9
   —————
    350.7

13.   341.17
    + 07.93
   —————
    449.00

MIXED Practice

Is each dotted line a line of symmetry? Write *yes* or *no*.

1. _____

2.  _____

3. 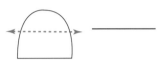 _____

Write the perimeter.

4. 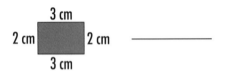 _____

5. 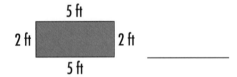 _____

Write the area in square centimeters. Each square is one square centimeter.

6.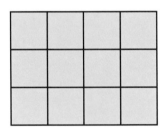

_____

7.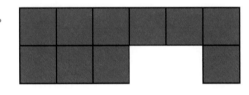

_____

Name _____

Subtracting Decimals

Subtract. Line up the decimal points.

| 1. | 12.43
− 8.24 | 2. | $76.15
− 44.49 | 3. | 91.83
− 37.69 |

4. 49.3 − 18.9   5. 27.48 − 12.8   6. $37.01 − $3.69

7. $67.17 − $58.81   8. 92.19 − 27   9. 48 − 7.35

Check to see if each problem is correct. If the difference is correct, write *yes* in the box. If it is incorrect, rework the problem to find the difference.

| 10. | 91.85
− 48.67
53.28 | 11. | $10.27
− 9.73
$1.54 | 12. | 413.61
− 87.97
325.64 | 13. | 341.17
− 129.88
124.75 |

 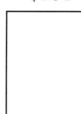  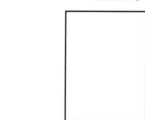

Solve. Remember to label your answers.

14. Grace has $35.00 to buy a rake and a shovel that cost $16.79 each. Does she have enough money?

   a. What do you need to do first to solve this problem?

   _____

   b. What do you need to do next? _____

   _____

   _____

   c. Does Grace have enough money?

   _____

15. Tally went to the grocery store to buy some lunch meat for her picnic. She bought 4.82 pounds of ham and 5.13 pounds of turkey. Which lunch meat is closer in weight to 5 pounds? Explain how you know.

   _____

   _____

MIXED Practice

Add or subtract. Show your work.

1. $37.05 - 18.96$

2. $3,509 + 6,387$

3. $7,025 - 3,519$

4. $46.95 + 35.67$

Name _____

Problem-Solving Strategy: Guessing and Checking

Solve. Remember to label your answers.

1. The sum of two numbers is 15. The difference between the two numbers is 1. Find the numbers.

   _____ and _____

2. The difference between two numbers is 2. The sum of the two numbers is 14. Find the numbers.

   _____ and _____

3. Angie made 2 sandwiches in the same amount of time that Gail made 1 sandwich. If they made 24 sandwiches altogether how many did each girl make?
   Angie _____
   Gail _____

4. Mrs. Jeffers told her son, "I'm thinking of two different numbers whose sum is 24. Both have 6 as a factor." What are the numbers?

   _____ and _____

5. Mary has 15 pennies. For every 3 pennies she has 1 nickel. How many nickels does she have? _____
   How much money does she have? _____

6. Isabel's mother is 4 times as old as Isabel. In 10 years, Isabel will be 18.
   How old is Isabel? _____
   How old is Isabel's mother?

   _____

7. Write your own problem similar to the ones on this page. Ask someone to solve it using the guess-and-check strategy.

   _____

   _____

   _____

Name _____

Shopping and Decimals

Use the prices at the department store to solve the problems.

1. You have $60.00. What three different items could you buy?

   a. List them. _____

   b. How much would your three items cost altogether?

   _____ + _____ + _____ = _____

   c. How much money would you have left? _____

2. How much would you spend if you bought two shirts and one pair

   of pants? _____

3. Which would cost more, two jackets or four pairs of shorts?

   _____

4. What two different items can you buy for exactly $51.53?

   _____

5. How much money would you spend if you bought one of every item?

   _____

Name _____

Chapter 12 Review

Write the decimal. Then shade to show each amount.

1. $\frac{7}{10}$ _____

2. $\frac{3}{10}$ _____

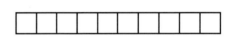

3. $1\frac{6}{10}$ _____

4. $2\frac{2}{10}$ _____

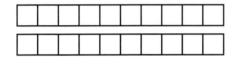

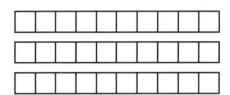

Write each fraction as a decimal.

5. $\frac{58}{100}$ = _____

6. $\frac{7}{100}$ = _____

7. $\frac{90}{100}$ = _____

8. $7\frac{82}{100}$ = _____

9. $19\frac{1}{100}$ = _____

10. $74\frac{34}{100}$ = _____

Solve.

11. If ✏ = 8 and ❖ = 6, what is (✏ × ❖) + ✏ ? _____

12. Darla's brother James is twice her age. In 20 years, they will be 7 years apart. How old are Darla and her brother right now?

_____

Compare. Write <, >, or =.

13. 13.28 ◯ 13.85

14. 6.6 ◯ 6.60

15. 12.32 ◯ 21.32

16. 83.92 ◯ 83.2

Order the decimals from greatest to least.

17. 10.6, 10.17, 10.28, 10.46   _____, _____, _____, _____

18. 41.16, 42.4, 42.34, 44.49   _____, _____, _____, _____

19. 75.43, 75.15, 75.3, 75.35   _____, _____, _____, _____

Add or subtract.

20. 45.82 + 25.93   21. 92.49 − 3.46   22. $15.92 + $69.35

23. 85.32 − 36.8   24. $72.20 − $29.48   25. 83.17 + 18

Solve. Remember to label your answers.

26. If Debra spends $18.24 on a lamp and $12.49 on a DVD, how much
     money does she spend in all?

     _____

27. If Debra pays for the items in problem 26 with four ten-dollar bills,
     how much change does she receive?

     _____

Name _____

Chapter 12 Test Prep

Ring the letter of the correct answer.

1. There are two mystery numbers whose product is 24. One number is even and the other number is odd. Which of the following could be the numbers?

   a. 2 and 12   **b.** 3 and 8   **c.** 4 and 6   **d.** none of these

2. Which decimal is the same as $\frac{21}{100}$?

   a. 2.1   **b.** 0.21   **c.** 0.021   **d.** none of these

3. Which decimal is the same as the mixed number $1\frac{4}{10}$?

   a. 1.04   **b.** 1.4   **c.** 14.0   **d.** none of these

4. If ❖ = 3 and ◕ is 5, what = (❖ × ◕) + ❖?

   a. 15   **b.** 18   **c.** 21   **d.** none of these

5. Compare. $\frac{1}{10}$ ◯ 0.1

   a. <   **b.** >   **c.** =   **d.** none of these

6. Compare. 1.74 ◯ 1.47

 a. <　　　　　b. >　　　　　c. =　　　　　d. none of these

7. Add $42.68 + $36.81.

 a. $80.59　　　b. $79.59　　　c. $78.49　　　d. none of these

8. Subtract 17.58 − 3.92.

 a. 14.46　　　b. 13.66　　　c. 13.46　　　d. none of these

9. If you buy 2 books at $7.83 each, how much will you spend in all?

 a. $15.66　　　b. $15.86　　　c. $16.66　　　d. none of these

10. If you spend $8.24, how much change will you receive if you pay with a twenty-dollar bill?

 a. $11.76　　　b. $11.86　　　c. $12.24　　　d. none of these

Name _____

Mental Math: Multiplying Multiples of 10, 100, and 1,000

Multiply.

1. 5×3 tens = _____ tens, so $5 \times 30 =$ _____, and $5 \times 300 =$ _____

2. 2×8 tens = _____ tens, so $2 \times 80 =$ _____, and $2 \times 800 =$ _____

3. 6×3 hundreds = _____ hundreds, so $6 \times 300 =$ _____

   and $6 \times 3,000 =$ _____

4. _____ $= 4 \times 90$

5. $8 \times 300 =$ _____

6. $6 \times 7,000 =$ _____

7. $3 \times 4,000 =$ _____

8. _____ $= 6 \times 400$

9. $7 \times 80 =$ _____

10. $6 \times 900 =$ _____

11. $5 \times 6,000 =$ _____

12. _____ $= 6 \times 800$

Solve. Remember to label your answers.

13. Marianne read 30 pages each day for 9 days. How many pages did she read in all?

14. Each bag of potato chips weighs 10 ounces. What is the weight of 8 bags of chips? 80 bags? 800 bags?

_____   _____

MIXED Practice

Solve.

1. $\begin{array}{r} 3,003 \\ -\ 1,321 \\ \hline \end{array}$

2. $\begin{array}{r} \$25.16 \\ +\ 34.73 \\ \hline \end{array}$

3. $\begin{array}{r} \$66.16 \\ -\ 37.07 \\ \hline \end{array}$

4. $3\overline{)16}$

5. _____ $= 9 \times 7$

6. $18 \div 6 =$ _____

Name _____

Multiplication Patterns

Look at the patterns.

$2 \times 3 = 6$		$4 \times 7 = 28$
$2 \times 30 = 60$	← 1 zero →	$4 \times 70 = 280$
$2 \times 300 = 600$	← 2 zeros →	$4 \times 700 = 2,800$
$2 \times 3,000 = 6,000$	← 3 zeros →	$4 \times 7,000 = 28,000$

Multiply and then write the correct number of zeros.

Multiply.

1. $5 \times 70 =$ _____
   THINK
   $5 \times 7 =$ _____
   Then write 1 zero.

2. $4 \times 60 =$ _____
   THINK
   $4 \times 6 =$ _____
   Then write 1 zero.

3. $5 \times 900 =$ _____
   THINK
   $5 \times 9 =$ _____
   Then write 2 zeros.

4. $4 \times 90 =$ _____

5. $3 \times 70 =$ _____

6. $8 \times 20 =$ _____

7. _____ $= 3 \times 400$

8. $7 \times 2,000 =$ _____

9. _____ $= 4 \times 80$

10. $6 \times 600 =$ _____

11. _____ $= 7 \times 70$

12. $5 \times 5,000 =$ _____

Extra Challenge

Follow the pattern above to complete.

13. $2 \times 40,000 =$ _____

14. $3 \times 90,000 =$ _____

15. $30 \times 40 =$ _____

16. $200 \times 40 =$ _____

Name _____

Estimating Products by Rounding

Estimate each product by rounding.

1. $3 \times 27 =$ _____
   THINK
   $3 \times 30 =$ _____

2. $4 \times 37 =$ _____
   THINK
   $4 \times 40 =$ _____

3. $5 \times 81 =$ _____
   THINK
   $5 \times 80 =$ _____

4. $3 \times 62 =$ _____

5. $5 \times 59 =$ _____

6. $8 \times 19 =$ _____

Estimate each product by rounding.

7. $\begin{array}{r} 75 \\ \times\ 6 \\ \hline \end{array}$

8. $\begin{array}{r} 83 \\ \times\ 4 \\ \hline \end{array}$

9. $\begin{array}{r} 31 \\ \times\ 8 \\ \hline \end{array}$

10. $\begin{array}{r} \$75 \\ \times\ 3 \\ \hline \end{array}$

11. $\begin{array}{r} 67 \\ \times\ 4 \\ \hline \end{array}$

12. $\begin{array}{r} 82 \\ \times\ 7 \\ \hline \end{array}$

13. $\begin{array}{r} 32 \\ \times\ 9 \\ \hline \end{array}$

14. $\begin{array}{r} \$68 \\ \times\ 3 \\ \hline \end{array}$

Ring the best estimate.

15. 5×62
    a. 30
    b. 300
    c. 3,000

16. 4×95
    a. 40
    b. 400
    c. 700

17. 6×37
    a. 300
    b. 240
    c. 180

Solve. Remember to label your answers.

18. Every day Jason does 75 push-ups. About how many push-ups does he do in 7 days?

19. How many minutes are in 9 hours?

_____ _____

Name _____

Multiplying, No Regrouping

Multiply.

1.	2.	3.	4.	5.
22 × 4	130 × 3	42 × 1	313 × 2	201 × 3

6.	7.	8.	9.	10.
30 × 2	22 × 3	202 × 4	322 × 2	14 × 2

11.	12.	13.	14.	15.
412 × 2	402 × 2	41 × 2	312 × 2	11 × 8

Solve. Remember to label your answers.

16. There are 3 rows of shopping carts outside a grocery store. Each row has 23 carts. How many carts are there altogether?

_____

17. Mr. Marks bought 4 dozen ears of corn. How many ears of corn did he buy?

_____

18. Cindy filled 21 pages of her photo album. There are 4 pictures on each page. How many pictures are there in all?

_____

19. There are 33 people on each bus. If there are 3 buses, how many people are there in all?

_____

Name _____

Multiplying 2-Digit Numbers with Regrouping

Multiply.

1.   15
   × 5

2.   36
   × 2

3.   45
   × 2

4.   18
   × 4

5.   24
   × 3

6.   28
   × 3

7.   15
   × 4

8.   27
   × 2

9.   22
   × 4

10.   19
    × 5

11.   14
    × 3

12.   38
    × 2

13.   29
    × 3

14.   39
    × 2

15.   18
    × 5

Solve. Remember to label your answers.

16. How many donuts are in 5 dozen?

17. How many inches are in 6 feet?

_____   _____

MIXED Practice

Solve.

1. $\frac{1}{3}$ of 18 = _____

2. _____ = $\frac{1}{2}$ of 16

Write the decimal.

3. $\frac{6}{10}$ _____

4. $4\frac{3}{10}$ _____

5. four tenths _____

Name _____

Multiplying 3-Digit Numbers with Regrouping

Multiply.

1. 782
    × 4

2. 830
    × 6

3. 129
    × 2

4. 649
    × 5

5. 339
    × 6

6. 253
    × 5

7. 424
    × 9

8. 508
    × 6

Rewrite the problems vertically. Multiply.

9. 783 × 4

10. 496 × 4

11. 939 × 3

   × ____   × ____   × ____

Solve. Remember to label your answers.

12. How many inches are in 9 feet?

13. Sam has 314 pages of baseball cards. Each page has 9 cards. How many baseball cards are in Sam's collection?

MIXED Practice

Divide. Use an R to show each remainder.

1. 5)37

2. 3)23

3. 4)33

4. 9)40

5. 6)27

Name _____

Hink Pinks

A **Hink Pink** is a riddle with a two-word rhyming answer.
For example, a wetlands jumper would be a **bog frog**.
Use your multiplication skills and the letter code to answer these
Hink Pinks.

A	C	D	E	F	H	I	L
470	1,242	3,220	189	192	544	3,455	2,536

M	O	R	S	T	V	Y	Z
1,548	2,524	168	738	405	440	1,776	260

1. This is a number to the rescue.

   $\overline{}$ $\overline{}$ $\overline{}$ $\overline{}$ $\overline{}$ $\overline{}$ $\overline{}$ $\overline{}$
   65×4 27×7 28×6 631×4 68×8 27×7 28×6 631×4

2. This is a piece of farm equipment that multiplies.

   $\overline{}$ $\overline{}$ $\overline{}$ $\overline{}$ $\overline{}$ $\overline{}$
   48×4 94×5 138×9 45×9 631×4 28×6

   $\overline{}$ $\overline{}$ $\overline{}$ $\overline{}$ $\overline{}$ $\overline{}$ $\overline{}$
   45×9 28×6 94×5 138×9 45×9 631×4 28×6

3. This is what you would call poetic products.

$\overline{45 \times 9}$ $\quad$ $\overline{691 \times 5}$ $\quad$ $\overline{258 \times 6}$ $\quad$ $\overline{27 \times 7}$ $\quad$ $\overline{123 \times 6}$

$\overline{28 \times 6}$ $\quad$ $\overline{68 \times 8}$ $\quad$ $\overline{888 \times 2}$ $\quad$ $\overline{258 \times 6}$ $\quad$ $\overline{27 \times 7}$ $\quad$ $\overline{123 \times 6}$

4. This is something fun to play on at the math playground.

$\overline{644 \times 5}$ $\quad$ $\overline{691 \times 5}$ $\quad$ $\overline{55 \times 8}$ $\quad$ $\overline{691 \times 5}$ $\quad$ $\overline{644 \times 5}$ $\quad$ $\overline{27 \times 7}$

$\overline{123 \times 6}$ $\quad$ $\overline{317 \times 8}$ $\quad$ $\overline{691 \times 5}$ $\quad$ $\overline{644 \times 5}$ $\quad$ $\overline{27 \times 7}$

Name _____

Problem Solving Application: Solving a Two-Step Problem

Solve. Use the bar graph.

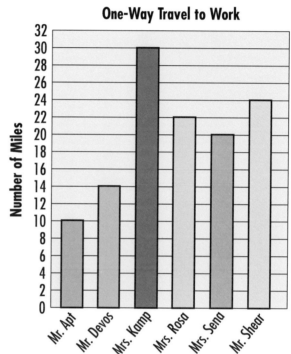

One-Way Travel to Work

1. How many miles does Mr. Apt drive to and from work each day?

   _____

2. How many miles does Mrs. Rosa drive to and from work in 9 days?

   _____

3. Who drives the greatest distance to work? How many miles does that person drive to and from work in five days?

   _____

4. How many more miles does Mrs. Sena drive to and from work each day than Mr. Devos?

   _____

5. Mr. Shear drove to work and then drove back home. Then he drove 6 miles to the grocery store, and drove 6 miles back home. How many miles did he travel that day?

   _____

Name _____

Dividing 2-Digit Numbers

Divide. Use long division.

	Steps to Remember when Dividing
	1. **D**ivide
	2. **M**ultiply
	3. **S**ubtract
	4. **C**ompare
	5. **B**ring down
	6. **R**epeat the steps

1. $9\overline{)99}$

2. $4\overline{)56}$

3. $7\overline{)84}$

4. $3\overline{)72}$

5. $6\overline{)84}$

6. $64 \div 4$

7. $58 \div 2$

8. $70 \div 5$

9. $81 \div 3$

10. $96 \div 6$

11. $98 \div 2$

12. $85 \div 5$

Name _____

Division Problems

Solve. Remember to label your answers.

1. Mr. Perry set up 98 chairs for a dance recital. He placed 7 chairs in each row. How many rows of chairs did he set up?

2. Cheryl glued 78 pictures in a scrapbook. She glued 6 pictures on each page. How many pages did she use?

3. One container can hold 3 tennis balls. How many containers are needed to hold 84 tennis balls?

4. Marlin bought 85 cookies to put in goody bags for his son's birthday party. He placed 5 cookies in each bag. How many goody bags did he fill?

5. Eli spent $84 on some books. If the cost of each book was $6, how many books did Eli buy?

6. Erin sorted her 96 collector cards into stacks of 8. When she finished sorting, how many stacks did she have?

Name _____

Chapter 13 Review

Multiply.

1. $6 \times 30 =$ _____

2. _____ $= 4 \times 7,000$

3. $6 \times 800 =$ _____

4. $\begin{array}{r} 11 \\ \times\ 7 \\ \hline \end{array}$

5. $\begin{array}{r} 23 \\ \times\ 3 \\ \hline \end{array}$

6. $\begin{array}{r} 12 \\ \times\ 4 \\ \hline \end{array}$

7. $\begin{array}{r} 33 \\ \times\ 3 \\ \hline \end{array}$

8. $\begin{array}{r} 13 \\ \times\ 5 \\ \hline \end{array}$

9. $\begin{array}{r} 28 \\ \times\ 2 \\ \hline \end{array}$

10. $\begin{array}{r} 28 \\ \times\ 3 \\ \hline \end{array}$

11. $\begin{array}{r} 16 \\ \times\ 4 \\ \hline \end{array}$

12. $\begin{array}{r} 463 \\ \times\ 5 \\ \hline \end{array}$

13. $\begin{array}{r} 509 \\ \times\ 6 \\ \hline \end{array}$

14. $\begin{array}{r} 832 \\ \times\ 7 \\ \hline \end{array}$

15. $\begin{array}{r} 703 \\ \times\ 4 \\ \hline \end{array}$

Estimate each product by rounding.

16. $\begin{array}{r} 67 \\ \times\ 4 \\ \hline \end{array}$

17. $\begin{array}{r} 83 \\ \times\ 4 \\ \hline \end{array}$

18. $\begin{array}{r} 59 \\ \times\ 5 \\ \hline \end{array}$

19. $\begin{array}{r} 72 \\ \times\ 4 \\ \hline \end{array}$

Solve. Remember to label your answers.

20. How many feet are in 15 yards?

21. One bag of marbles contains 75 marbles. How many marbles are in 5 bags?

_____   _____

Divide. Use long division.

22. 4)60   **23.** 3)87   **24.** 5)85   **25.** 7)91

Solve. Remember to label your answers.

26. Devon built 45 model cars. He put an equal number of cars on each of 3 shelves. How many cars did he put on each shelf?

_____

27. Francine was collecting toys to give away to charity. She collected 45 toys in one box. She collected 2 bags each with 9 toys. How many toys did Francine collect in all?

_____

Name _____

Chapter 13 Test Prep

Ring the letter of the correct answer.

1. If $5 \times 4 = 20$, $5 \times 400 =$ _____

    a. 20   **b.** 200   **c.** 2,000   **d.** 20,000

2. $8 \times 8,000 =$ _____

    a. 6,400   **b.** 64,000   **c.** 640,000   **d.** 6,400,000

3. Estimate. 48×4

    a. 160   **b.** 200   **c.** 240   **d.** 400

4. Multiply. 61×9

    a. 540   **b.** 549   **c.** 550   **d.** 610

5. Multiply. $28 \times \$8$

    a. $224   **b.** $232   **c.** $240   **d.** $256

6. Multiply. 547×6

   a. 3,000   b. 3,142   c. 3,199   d. 3,282

7. Divide. $72 \div 3$

   a. 22   b. 23   c. 24   d. 25

8. Divide. $96 \div 6$

   a. 11   b. 16   c. 26   d. 61

9. Sheets were on sale for $28 per set. Sloane bought 3 sets for her new room. How much money did she spend?

   a. $75   b. $81   c. $84   d. $94

10. Tori bought a baseball glove for $14.98 and three baseballs at $8.88 each. How much did she spend in all?

    a. $45.00   b. $41.62   c. $39.39   d. $35.35